CISM COURSES AND LECTURES

The series presents lecture notes, monographs, edited works and proceedings in the field of Mechanics, Engineering, Computer Science and Applied Mathematics.
Purpose of the series in to make known in the international scientific and technical community results obtained in some of the activities organized by CISM, the International Centre for Mechanical Sciences.

Proceedings of the ISSEK94 Workshop on

MATHEMATICAL AND STATISTICAL METHODS IN ARTIFICIAL INTELLIGENCE

EDITED BY

G. DELLA RICCIA
UNIVERSITY OF UDINE

AND

R. KRUSE
UNIVERSITY OF BRAUNSCHWEIG

AND

R. VIERTL
TECHNICAL UNIVERSITY OF WIEN

SPRINGER-VERLAG WIEN GMBH

Le spese di stampa di questo volume sono in parte coperte da
contributi del Consiglio Nazionale delle Ricerche.

This volume contains 32 illustrations

In order to make this volume available as economically and as
rapidly as possible the authors' typescripts have been
reproduced in their original forms. This method unfortunately
has its typographical limitations but it is hoped that they in no
way distract the reader.

ISBN 978-3-211-82713-0   ISBN 978-3-7091-2690-5 (eBook)
DOI 10.1007/978-3-7091-2690-5

PREFACE

This volume contains the papers accepted for presentation at the invitational ISSEK94 Workshop on "Mathematical and Statistical Methods in Artificial Intelligence" organized by the International School for the Synthesis of Expert Knowledge (ISSEK) and held at the Centre International des Sciences Mécaniques (CISM) in Udine from September 6 to 8, 1994.

In recent years it has become apparent that an important part of the theory of Artificial Intelligence is concerned with reasoning on the basis of uncertain, incomplete or inconsistent information. Classical logic and probability theory are only partially adequate for this, and a variety of other formalisms have been developed, some of the most important being fuzzy methods, possibility theory, belief function theory, non monotonic logics and modal logics.

The aim of this workshop was to contribute to the elucidation of similarities and differences between the formalisms mentioned above. The talks were given by researchers that are well known in their respective field and the discussion was focused on such topics as fuzzy data analysis, probabilistic reasoning, learning and abduction and logics in uncertainty. Moreover there was an interesting session on industrial applications with talks given by members of ISSEK. This workshop was very successful and led to joint research and fruitful collaboration.

We would like to thank the following Institutions: the Universities of Braunschweig and Wien for their scientific support in the preparation of the program, CISM for hosting the meeting and the University of Udine for providing technical support.

This workshop was made possible by a grant from the Fondazione Cassa di Risparmio di Udine e Pordenone to which we express our gratitude.

The next invitational ISSEK Workshop will take place at Udine in 1996.

R. Kruse

CONTENTS

Page

FUZZY SPECIFICATION OF UNCERTAIN KNOWLEDGE AND VAGUE OR IMPRECISE INFORMATION

H. Bandemer
TU Bergakademie Freiberg, Freiberg, Germany

ABSTRACT

Impreciseness and vagueness are facets of uncertainty and mean that an object or some of its features can not be recorded or described precisely. Hence both must be considered in contrast with *randomness*, which describes variability, being another kind of uncertainty. Whereas probability theory and mathematical statistics deal with the behaviour of (perhaps hypothetical) populations, impreciseness and vagueness are concerned with each single piece of information, called a *datum*. For handling such data they must be described as mathematical items. The "classical" model for *impreciseness* is given by *set-value-description*, e.g. by intervals in the simplest case. However, as is known, the main problem in application of interval mathematics is fixing *precise* ends of the intervals. Moreover, data can be given by *verbal descriptions*. Then they are called *vague* data, usually coded by numbers or letters. However, in such a form they lose much of their semantic meaning, which would be very important for processing them and for interpreting the conclusions drawn from this processing. In the paper some examples are provided how such imprecise or vague data had been specified by fuzzy sets in real-world applications, e.g. measuring from blurred pictures in the micro area and comparing imprecise sample-spectrograms with standard spectrograms. Methods included are, e.g., using grey-tone levels in pictures on the screen and using structure elements from mathematical morphology for specifying fuzzy functions. Finally a numerical example is considered to show how impreciseness influences the results of simple statistical procedures.

1. Introduction

The wording "datum" means, literally, "something actually given". It makes sense only in a certain context and expresses that "something" was found in a state characterized by just this datum. Obviously, such a datum contains information only if there are at least two different possibilities for the state of the "something" in question. Hence we can consider every datum as a realisation of a certain *variable* in a suitable set of values, called the *universe of discourse*, and reflecting these possibilities for the state in the given context.

The first problem in a mathematical modelling of such an affair consists in a mathematical representation of the possible data simultaneously specifying a suitable universe. Usually the data are already given in some mathematical form. Here we will mention some examples by catch words: qualitative properties (binary); grades, shades, nuances (marks, natural numbers); observations and measurements (real numbers or vectors); pictures; statements of experts (wordings).

When considering these data more and more intensively we find them all burden with *uncertainty* of different kinds. Hence conclusions derived from them will inherit this uncertainty. Possibly the uncertainty is even increased by the method of analysis applied. Ignoring these circumstances the conclusions can become worthless or even misleading. Uncertainty of data can occur in different forms: variability, impreciseness and vagueness, and has different forms; (see e.g. [1]).

First we will consider results of observation and measurement and distinguish two principal cases.

1. The results of a measurement or observation procedure are assumed to be *precise*, however, vary when the procedure is performed repeatedly. In this case we usually take the *model of randomness*. We imagine a more or less theoretic population of results and the actual value obtained is explained to be chosen *at random* according to some probability law. The appropriate mathematical theory for handling such problems and inference is supplied by *stochastics*: probability theory and mathematical statistics.

   Another aspect of randomness is *belief*. Probability is not the only model for belief, since additivity is sometimes lacking. Here the notion of a fuzzy measure is a suitable model. (e.g. [24])

2. The other principal case occurs if there is *no unique value* describing the actual piece of information given by the observation or measurement procedure, the so-called case of *impreciseness*. If there is available some set of values able to describe the information, then usually *interval mathematics* (or some formal analogue) are used for handling such problems of modelling and inference. The cruxial problem of this approach is the specification of the *exact* set of elements, e.g. the ends of the interval. If the data are given by verbal description, then they are called

vague. This is the typical case with prior knowledge or side information. When representing them by numbers or letters such data lose much of their semantic meaning, which would be important for processing them and for interpreting the conclusions drawn from this processing.

Usually all these kinds of uncertainty, randomness, impreciseness, and vagueness, occur simultaneously in the same situation. As an example we mention the catching of an animal in the wilderness for inspection. The animal is chosen by chance out of a population of the species (randomness), measured with respect to weight and length with expedition tools (impreciseness), and evaluated by a doctor with respect to health, e.g. fatness and infestation (vagueness).

Whereas probability theory and mathematical statistics deal with the behaviour of (perhaps hypothetical) populations, impreciseness and vagueness are concerned with each single piece of information, which was called a *datum* in the beginning of this section.

When trying to specify uncertainty, given by impreciseness or vagueness we will first consider some possible reasons for such uncertainty.

When considering the temperature of some room we know that there is always a temperature field spread out within the room, where the values vary in place and time. Hence, when observing a given thermometer we see only the temperature value in only one place and its environment. The precision of the thermometer is only with respect to the value at the place of it and has nothing to do with the field characterising the temperature in the *whole room*. On the other hand, the value read from the thermometer will usually suffice, e.g. for controlling the heating for the room. Hence this datum represents the state of the object only *symbolically*.

This fact can be expressed by specifying the temperature of the room by a *fuzzy set*.

The main idea of fuzzy set theory is the permission that some element x of the universe U can have a *gradual* membership $m_A(x)$ to the set A, usually a value taken from the closed interval $[0, 1]$. In such a way a function m_A is defined, called *membership function*,

$$m_A \mid U \to [0, 1]. \tag{1.1}$$

With respect to the above example we can interpret the value $m_A(x)$ as the *degree of acceptance* that x characterises the temperature of the whole room adequately. Hence we will start with the observed "pseudo-exact" value and endow it with an interval of evaluated values of decreasing acceptance.

Another case, leading to a similar specification is given, when there is a precise value representing the state of an object, but the sensor is *coarse*. Such a case occurs, e.g., when measuring length and weight under household conditions (a spoonful, a pinch), the measuring of an animal with expedition tools (see above), or estimating the age of an animal.

According to our experience such a observational fuzziness is frequently much larger than error variance of measurements. Omitting of observational fuzziness leads to euphoric impression of measurement and result preciseness.

It is possible to model randomness and fuzziness simultaneously (see [13], [14], [21], [22], [23]).

2.   Specification of fuzzy sets

Sometimes the question is raised, whether there is a method to specify membership functions *uniquely* from the practical problem, like a differential equation from a mechanical problem. When this question is answered in the negative, this fact is used by opponents of fuzzy set theory for blaming the theory for this supposed drawback. From the standpoint of an engineer the numerical assessment of a membership function is a modelling of second or even third order to take into account effects of uncertainty. Demanding precise specifications would contradict the whole concept of fuzzy thinking in daily life as well as in applied sciences. Nobody would in earnest trouble an engineer with the question, how he has specified the coefficients and boundary values of his differential equation, of which he assumes that it is adequate with the problem he considers. Usually not even the structure of the differential equation is questioned. When using methods of mathematical statistics most frequently the data are assumed to be realisations of independent random variables with Gaussian distributions. In a concrete case this assumption is checked, if at all, by a quick glance to some critical value, which makes sense on certain other assumptions.

As our experience show, the conclusions drawn by methods of fuzzy set theory are only little dependent on the analytical form and the local values of the specified membership functions, provided that a certain local monoticity is kept.

Before starting with specifications of data by fuzzy sets we want to stress once more the fact that we will model *realisations* of variables and *not populations*, as is the case in probability theory.

For a more detailed consideration of these problems we refer to [4] and [9].

2.1   Parametric specification

The usual starting point will be a *pseudo-exact* datum, given as a "real number" and resulting from an observation or a measurement by a device of given sensitivity. Then "local fuzziness" is considered as a generalisation of the usual tolerance expression characterised by some Δx, as is known from the law of error transmission. Now a non-increasing function, say $h(d(x, z))$, where d is some distance, is chosen with $h(0) = 1$, containing some parameters for tuning.

Then, looking at the given pseudo-exact datum x the question is to be answered, which other values z are possible as observations with *the same degree* of acceptance as the realised datum x. These values form the *kernel* of the fuzzy datum to be specified. Next all the values are to be determined, which are taken to be *totally impossible* as observations, when x was observed. This is the complement of the *support* of the fuzzy datum. Then we have to bridge the zone between the kernel and this complement by a suitable monotonic function h, as simple as possible and as expressible as necessary. If the universe is a segment of the straight line, such fuzzy data are called *fuzzy numbers*. Usually h is chosen as a linear or a quadratic function, sometimes hyperbolic or even Gaussian.

An interesting approach is due to [15]. They choose a suitable *fuzzy equality relation* R expressing the relation "approximately equal"

$$m_R(x, z) = [1 - a \mid x - z \mid]^+, \tag{2.1}$$

and specify with this a fuzzy singleton, which leads to the usual hat function. i. e. with a linear h.

Considering multidimensional data or combinations of data ("samples") $A_i; i = 1, ..., l$ the problem arises, how to combine fuzzy numbers to obtain a fuzzy vector. As usual this can be done by forming a Cartesian product:

$$A = A_1 \times A_2 \times ... \times A_l. \tag{2.2}$$

In dependence on the chosen connection realising $\times$ by a suitable t-norm and the type of the membership functions for the fuzzy numbers different multidimensional types of "fuzzy points" are obtained:

$$m_A(x, y) = t(m_1(x), m_2(x)). \tag{2.3}$$

We mention two important examples:

$$A_1, A_2 \text{ linear}; t(u, v) = \min\{u, v\} : \text{tetraeder} \tag{2.4}$$

(see Figure 1) and

$$A_1, A_2 \text{ quadratic}; t(u, v) = \max\{u + v - 1, 0\} : \text{beans} \tag{2.5}$$

(see Figure 2).

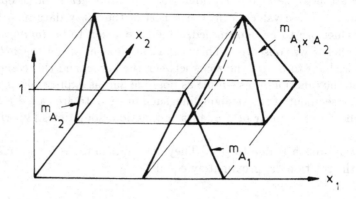

Fig. 1   Fuzzy point with tetraeder-formed membership function built as Cartesian product of two hat-shaped fuzzy numbers (cp. Formula (2.4))

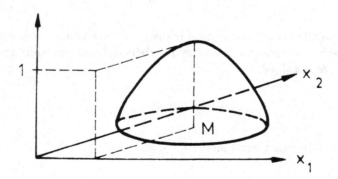

Fig. 2   Fuzzy point with bean-shaped membership function (cp. Formula (2.5))

Note that the bean-components are interactive, with respect to the min-product, however, they are very useful when considering linear transformation, e.g. for fuzzy partial least squares methods.

An other interesting suggestion is due to [11] and used by him in numerous successful practical applications. He uses the well-known Airy potential function

$$m(x) = \left(1 + \left(\frac{1}{b} - 1\right)\left(\frac{|x - x_0|}{c}\right)^d\right)^{-1}$$ (2.6)

where x_0, b, c, d are parameters for tuning. For connecting the fuzzy numbers to fuzzy vectors he suggests realising the Cartesian product by multiplying the membership functions with each other.

2.2 Specifying from the context directly

Fuzzy sets can be represented by grey-tone pictures, where the shade of grey reflects the membership value of the point. Sometimes grey-tone pictures represent the fuzziness of an observation or measurement.

As examples we consider observations under extreme conditions, when the object to be observed is extremely small or distant. A first case is the measuring of VICKER's hardness on a micro rectangular solid specimen, which was subjected to a hardening treatment onto one of its faces. Then the specimen was cut up orthogonally to this treated face. The inner plane produced in such manner was tested pointwise with respect to hardness, in order to obtain a *functional relationship between hardness* and the distance to the border, where the hardening treatment was applied, called the *depth*. The specimen was very small, caused by the intended application of the results, and for a high resolving power with respect to position, both, pressing power and impress, must be very small. The observation must be performed by presenting the specimen plane to a microscope and by enlarging the picture by means of an image processing equipment. The result of the observation was presented as a grey-tone picture of the impress on the screen (see Figure 3). This picture was influenced by sources for inaccuracy and imprecision, which could not be controlled by the experimenter and which essentially effect the vagueness of the result. So, lightness and contrast of the equipment were controlled automatically according to some system unknown to the experimenter. We mention further the human eye as a measuring tool, the light-optical lower bound of resolving power, wavelength of light, adjusting conditions at the microscope, scanning quality of the television equipment. Note, that a sharpening of edges in the grey-tone picture by means of mathematical morphology would raise precision only virtually. Moreover, the coordinates of the points, where hardness was measured on the specimen, are likewise subject to inaccuracy and uncertainty. The impress has a finite extent and thus measures hardness in a certain neighbourhood of the top of the pyramid. The screen showed always only a part of the plane under investigation and turning from segment to segment causes increasing inaccarucy of depth.

Fig. 3 VICKERS-impress in a specimen face taken from the screen

Hence both, hardness and depth, were modelled by fuzzy sets.
The details are described in [8] and [6].
Hence, only a sketch of the performance will be given. VICKER's hardness is defined
as the quotient of the pressing power p and the extent of the impress surface, say s,

$$h(p, s) = p/s \, . \tag{2.7}$$

Let d denote the length of the diagonal of the square base of the impress, then we have

$$s = d^2/c_0 \, , \tag{2.8}$$

where c_0 depends on the face vertex angle of the pyramid causing the impress when
pressed with power p onto the surface. Figure 3 shows a segment with impresses, which
are unintentionally sharpened, when the photo was taken from the screen.
The grey-tone pictures of impresses are interpreted as fuzzy sets G_j on R^2 where the
shades reflect the vagueness of the experimental result. The diagonals of this fuzzy
squares have the same directions as the axes of the chosen coordinate system. For
measuring the length of the diagonals the fuzzy regions G_j are transformed into their
corresponding *fuzzy contours*

$$C_j : m_{C_j}(x, y) = 2 \min\{m_{G_j}(x, y), 1 - m_{G_j}(x, y)\} \, . \tag{2.9}$$

Let y_{0j} be the coordinate of the *horizontal* diagonal of C_j. The corresponding membership function $m_{C_j}(x, y_{0j})$ splits into two parts which can each be approximated by

a fuzzy number, say M_{l_j} and M_{r_j}, respectively. Their difference

$$D_j = M_{r_j} \ominus M_{l_j} \qquad (2.10)$$

defines the *fuzzy length of the diagnal*, the membership function of which is to be calculated by

$$m_{D_j}(d) = \sup_{(x,z):d=x-z} \{m_{rj}(x),, m_{lj}(z)\}. \qquad (2.11)$$

An inspection of the α-level sets of D_j showed that an approximation of m_{D_j} by functions of triangular shape was appropriate.

The result of this modelling is a fuzzy number D_x representing the diagonal length, where, instead of the index j, the dependence on the depth x is inserted where the top of the pyramid touched the plane.

Omitting c_0, for the moment and convenience, the *fuzzy hardness* $H(x)$ at the *crisp* depth x can be computed from the given fuzzy length D_x by the extension principle

$$m_{H(x)}(h) = \sup_{z:h=p/z^2} m_{Dx}(z) = m_{Dx}((p/h)^{1/2}). \qquad (2.12)$$

For modelling the fuzziness of the measurement of depth a fuzzy number $X(x)$ with a linear symmetric membership function seems to be appropriate.

Since there is no reason to assume an interaction between the measurements of hardness and of depth, we combine the two fuzzy sets by the minimum operator

$$m_{H(X)}(h, z) = \min\{m_{H(x)}(h), m_{X(x)}(z)\}. \qquad (2.13)$$

To evaluate a functional relationship between hardness and depth the fuzzy observations $H_j(X)$ are aggregated by union

$$m_H(h, x) = \max_j m_{H_j(X)}(h, x). \qquad (2.14)$$

For a first impression we consider the modal trace of $m_H(h, x)$, i.e.,

$$H_F(x) = \{(h, x) \in R^2 : m_H(h, x) = \sup_u m_H(u, x) > 0\}. \qquad (2.15)$$

The modal trace is then used to specify a functional relationship, making sense in the practical context, resulting in

$$h(x; c, b, v, q) = c + b \exp\{-(x/v)^q/2\}, \qquad (2.16)$$

where c represents the hardness of the kernel of the material (not influenced by the hardening treatment), b is the maximal hardness increase, and v and q explain where and how fast hardness decreases with increasing depth. The result of the inference is shown in Figure 4, the graph of the most likely functional relationship for the hardness in dependence on the depth surronded by a grey-tone belt, induced by the fuzziness of the observation of hardness and depth.

Similar results of fuzzy inference from grey-tone pictures are obtained by [12] with respect to contours of distant flying objects, and by BANDEMER/KRAUT (up to now unpublished) with small veins in the background of a human eye, to investigate the effect of some medicine to increase the blood supply of the eye.

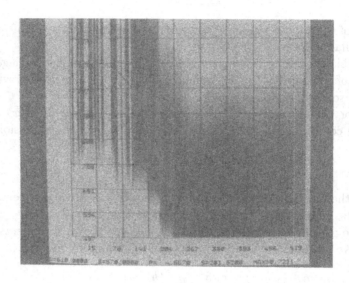

Fig. 4   The membership function of the aggregated fuzzy hardness data (grey shaded surface) the modal trace (lighter grey surface within the former one), and the "most likely" graph (the curve) of the functional relationship, as presented by the screen.

As a second case we consider indirecte observations, and as an example projections of three dimensional objects into the plane, e.g. X-ray photographs of tumours or projections of transparent tissues under normal light. Here the problem is the contour of the object. All points which belong surely to the object are transformed into black ones, independently of the original colour or grey-tone value; all points which belong surely not to the object become white. The points between get gradually grey values reflecting their fuzziness coming from the real three-dimensionality of the object. This grey-tone values reflect the membership to the object. For our purpose we need, however, the membership to the *contour* of the object. This is managed by intersection of the fuzzy set with its complement in analogy with the crisp case, where the contour is defined as the intersection of the closed set with its closed complement. This performance was used already in the example above (cp. (2.9)). In mineral processing the contour of a particle is decribed only fuzzily, e.g. by evaluating the deviation of it from a standard form, say a ball. For defining a fuzzy *roundness* the fuzzy contour, obtained e.g. according to the above performance, can be fitted by a functional relationship (see [5]) or by measuring the fuzzy area and the fuzzy perimeter of the fuzzy contour and using the given definition of roundness to define fuzzy roundness (see [7] and [9]).

2.3   Nonparametric specification from pseudo-exact data

Frequently pseudo-exact data are given without any hint to their local fuzziness. Even in this case problems of inference from these data can be solved taking into account inherent fuzziness of the data.

A possible approach uses means from *mathematical morphology*, see [20]. This theory is concerned with local transformations of binary pictures and aims at a cognition of the inner structure of picture patterns and scene recognition. A basic performance consists in the definition of a structural element, being much smaller than the picture to be investigated, and shifting this element across the picture. Then the value of the picture in the actual midpoint of the structural element is changed according to some law depending on all the values of the picture within the actual position of the structural element. For instance, the new value is *black*, if the structural element covers *only black* points of the picture: *erosion*, the new value is *black*, if the structural element covers *at least one black* point of the picture: *dilation*. Besides, the approach was generalised to functions, where erosion and dilation are now connected with minimisation and maximisation.

The idea of a structural element was adopted to specify fuzzy sets from given sets of pseudo-exact data ([9]).

Given a point cloud (see Figure 5) with unknown kind of local fuzziness.

A small disc was shifted across the set and the relative number of data hit by it was taken as the membership value of its actual midpoint. The aim was to find a general (fuzzy) trend. The result is a grey-tone picture, which looks like a natural range of mountains expressed by its level lines in a topographical map. The procedure was performed several times with different radii of the disc. When the radii were chosen too small, then most of the points were surrounded with disjoint small discs (a picture like a pea soup). When the radii were chosen too large, the procedure resulted in a big grey region (like a thundercloud). In both these cases no reasonable trend could be recognized. However, when the radius was chosen near the probable reach of impreciseness, then a reasonable trend became visible. An example is given in Figure 6.

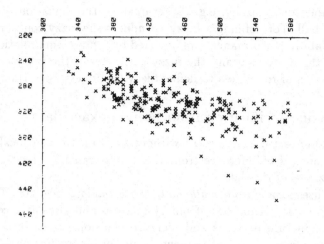

Fig. 5   Pseudo-exact data from histophotometrical measurements

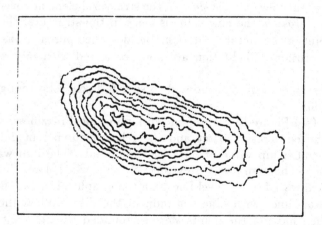

Fig. 6   Level lines for $\alpha = 0.10(0.05)0.45$ of the aggregated fuzzy observation set
specified from the data shown in Figure 5

A suitable approximation of the fuzzy trend is obtained by forming the *modal trace*

$$f_F(x) = argsup_y m(x,y), \tag{2.17}$$

where the membership function $m(x,y)$ is the grey-tone level at the point (x,y). A
picture of $f_F(x)$ is shown in Figure 7.

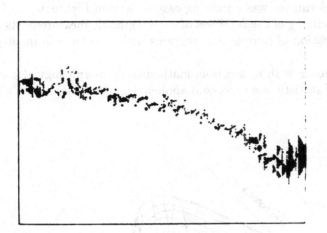

Fig. 7   Modal trace of the fuzzy observation set specified from the data shown
          in Figure 5

In general the modal trace is neither unique nor continuous, but will show the behaviour
of the ridge of the corresponding mountain range. For details see [10].

For an other example consider a bundle of trajectories, as shown in Figure 8.

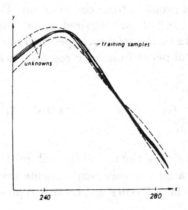

Fig. 8   Absorbance spectra (absorbance y versus wavelength x) for dissolved
          pain relieving tablets as a fingerprint for spectroscopic quality control

For every value of x a small segment of a straight was shifted in parallel to the y-axis
and the relative number of trajectories hit was taken as the membership value of its

actual position. The staircase function was approximated by a quadratic function. The result of this performance was a running cage as seen in Figure 9.

The aim was specifying of the fuzziness of spectrographic measurements being used for a matching evaluation of sample spectrograms with standards in quality control. For details see [18].

Besides this approach with means from mathematical morphology also methods based on mathematical statistic are successful applied.

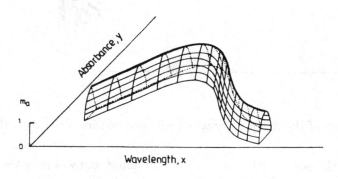

Fig. 9   Example of a UV-spectrum considered as a fuzzified sample spectrum

As an example we mention a problem from conservation. For an automatic ecological device a simple and robust method was required using reference and sample spectra for pollulants. Here the peaks of the reference spectrogram were surrounded by fuzzy beans approximating clouds of peaks from spectrograms of a given data base

$$
\begin{aligned}
B_i : m_i(x,y;x_i,y_i) = \\
[1 - c_1(x - x_i)^2 - c_2(y - y_i)^2 - c_{12}(x - x_i)(y - y_i)]^+ ; \\
i = 1, \ldots, M ,
\end{aligned}
\tag{2.18}
$$

where (x_i, y_i) are the coordinates of the top of the i-th peak of the reference spectrum an c_1, c_2, c_{12} are determined after extensive experiments using an available data base and experience of experts. The *reference spectrum* is represented by the union of all the beans as given in (2.18)

$$
B = \bigcup_{i=1}^{M} B_i .
\tag{2.19}
$$

Then the *crisp spectrum*

$$A = \{(x_{A_1}, y_{A_1}), \ldots, (x_{A_N}, y_{A_N})\} \tag{2.20}$$

is recorded. For eliminating systematic deviations to *all* the peaks in retention time (x-axis) and intensity (y-axis) the sample spectrum is shifted within pre-assigned small intervalls Δx and Δy, respectively, in both these directions. Now the similarity relation S is specified by

$$m_S(A, B) = \sup_{\substack{\Delta x \\ \Delta y}} \mathsf{card}(A \cap B)/\mathsf{card} A, \tag{2.21}$$

which means in full detail

$$m_S = \sup_{\substack{|\varepsilon_x| \leq \Delta x \\ |\varepsilon_y| \leq \Delta y}} \sum_{j=1}^{N} m_i(x_{A_j} + \varepsilon_x, y_{A_j} + \varepsilon_y; x_i, y_i)/N. \tag{2.22}$$

An example is shown in Figure 10, where the beans are outlined as ellipses.

Giving an upper bound, say m_0, for m_s, the similarity relation m_s is used in pratice to sort out sample spectra automatically, which are "far from" the reference spectrum. An automatic ecological supervision device being sensible to some given pollutant spectra, stored in the memory, will alarm, if m_s is higher than m_0.. Here m_0 is chosen after some training run to avoid too many false alarms as well as an overlooking of any dangerous situation (see [17]; see Figure 10).

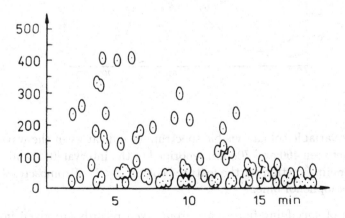

Fig. 10 Fuzzy peaks obtained from a capillary chromatogram

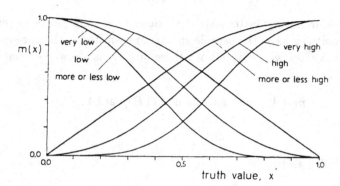

Fig. 11 Membership function for characterizing "solubility in water"

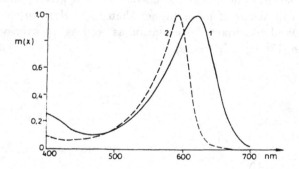

Fig. 12 Feature variable colour: visible spectrum of indicators in the wavelength
range between 400 and 700 nm renormed to the interval $[0, 1]$ and used as
membership function for the feature "colour".   1 – Bromoscresol green,
2 – Bromophenol blue.

Other examples of specifying fuzzy data from given records are given in the context
of recognition of contours in mineral deposits ([2]) and identifying ancestors from genealogical records, where names and dates are fuzzy ([3]).

Finally, even verbal given data are expressed by fuzzy sets, so e.g. in a chemometric context, solubility (see Figure 11) and colours of chemicals (see Figure 12). For details see [19].

Usually, when handling colours, it is suggested to define several features in parallel evaluating different ground colours, in such manner specifying a lattice structure for the colours. In the special case this seemed to be unnecessary: The visible spectrum was renormed to evaluate the colour. The approoach was used to predict the colour of a not yet produced compound successfully (see [19]).

3. Example with fuzziness and randomness

In statistical inference usually observational fuzziness is omitted in favour of random variability. Especially, if the data are subjected to numerical procedures, even small local fuzziness can effect high fuzziness of the result. A usual method from mathematical statistics is regression analysis. Frequently the design matrices have weak condition, however, this is usually not reflected within the result.

In the following a simple example from a textbook ([16]) is reconsidered.

From French national accounts for the years 1949 to 1959 figures for certain global quantities, namely imports, gross domestic production, stock formation and consumption are extracted. These quantities are expressed in millards of new francs at 1956 prices, and the results are shown in Table 1 and Figure 13.

Year	t	x_1 Imports	x_2 Gross domestic production	x_3 Stock-formation	x_4 Consumption
1949	1	12.6	117.0	3.1	84.5
1950	2	13.1	126.3	3.6	89.7
1951	3	15.1	134.4	2.3	96.2
1952	4	15.1	137.5	2.3	99.1
1953	5	14.9	141.7	0.9	103.2
1954	6	16.1	149.4	2.1	107.5
1955	7	17.9	158.4	1.5	114.1
1956	8	21.0	166.5	3.8	120.4
1957	9	22.3	177.1	3.6	126.8
1958	10	21.9	179.8	4.1	127.2
1959	11	21.0	183.8	1.9	128.7

Table 1 Imports, production, stock-formation and consumption in France

With the regression set-up

$$y = b + a_1 x_1 + a_2 x_2 + a_3 x_3 \tag{3.1}$$

and the 11 observations a fitting by the method of least squares was performed, resulting in the estimations for the coefficients

$$b = -8.67; a_1 = -.0; a_2 = 0.67; a_3 = 0.37 \tag{3.2}$$

and the observation value $y_{11} = 21$ was estimated by $y_{11} = 20.86$.

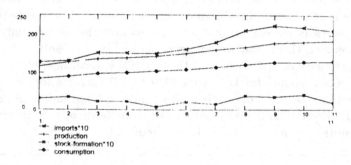

- imports*10
- production
- stock formation*10
- consumption

Fig. 13 Graphical representation of the data given in Table 1

Now, the data are assumed as burden with observational fuzziness by recording and rounding. Hence the given numbers are transfered into fuzzy numbers with linear membership functions choosing the corresponding spreads: 0.1 for y, x_1, x_3 and = .01 for x_2. Then the usual formula for linear exstimation by the method of least squares was extended by an extension principle using the Lukasiewicz-t-norm (the bounded product) for intersection

$$t(u, v) = \max\{0, u + v - 1\} \tag{3.3}$$

producing fuzzy estimates for the coefficients and the "true" values of the function. Figure 14 shows the estimation by the usual method of least squares and Figure 15 its fuzzy counterpart, where the grey-tones reflect the values of the membership function of the estimate.

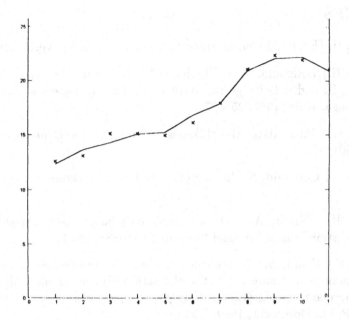

Fig. 14 Approximation of the data from Figure 13 by the method of least squares
using the set-up according to (3.1)

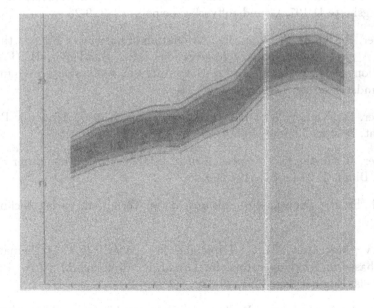

Fig. 15 Fuzzy grey-tone belt showing the influence of impreciseness of the data
on the approximation

REFERENCES

1. Bandemer, H. (Ed.) : Modelling Uncertain Data. Akademie Verlag, Berlin 1993

2. Bandemer, H.; Bellmann, A. : Unscharfe Methoden der Mehrphasenregression. In: Peschel, G. (Ed.): Beiträge zur Mathematischen Geologie und Geoinformatik, Sven von Loga, Köln, 1991, 25 - 27.

3. Bandemer, H. : Fuzzy data: the "Likeness" of individuals. Genealogical Computing 10 (1990), 47.

4. Bandemer, H.; Gottwald, S.: Introduction to Fuzzy Methods. Wiley, Chichester, 1995

5. Bandemer, H. ; Kraut, A. : On a fuzzy-theory-based computer-aided particle shape description. Fuzzy Sets and Systems 27 (1988), 105-113

6. Bandemer, H. ; Kraut, A. : A case study on modelling impreciseness and vagueness of observations to evaluate a functional relationship. In: Janko, M. ; Roubens, M. ; Zimmermann, H.-J., (Eds.): Progress in Fuzzy Sets and Systems, Kluwer Academic Publ., Dordrecht, 1990, 7-21

7. Bandemer, H. ; Kraut, A. : On fuzzy shape factors for fuzzy shapes. In: Bandemer, H. (Ed.): Some Applications of Fuzzy Set Theory in Data Analysis II, Freiberger Forschungshefte D 197, Grundstoffverlag Leipzig, 1989, 9-26

8. Bandemer, H.; Kraut, A.; Vogt, F. : Evaluation of hardness curves at thin surface layers - A case study on fuzzy observations - In: Bandemer, H. (Ed.): Some Applications of Fuzzy Set Theory in Data Analysis; Freiberger Forschungshefte D 187, Grundstoffverlag, Leipzig 1988, 9-26

9. Bandemer, H.; Näther, W. : Fuzzy Data Analysis. Kluwer Academic Publishers, Dordrecht, Boston, London 1992.

10. Bandemer, H.; Roth, K. : A method of fuzzy-theory-based computer-aided data analysis. Biom. J. 29 (1987), 497-504

11. Bocklisch, St. F. : Prozeanalyse mit unscharfen Verfahren. Verlag Technik, Berlin 1987

12. DiGesù, V.; Maccarone, M. C.; Tripiciano, M. : MMFUZZY: Mathematical morphology based on fuzzy operators. In: Lowen, R.; Roubens, M. (Eds.) Proc. IFSA 91, Brussels, Engineering, 1991, 29-32.

13. Goodman, I. R. ; Nguyen, H. T. : Uncertainty Models for Knowledge-Based Systems. North-Holland Publ. Comp., Amsterdam 1985

14. Kruse, R. ; Meyer, K.D. : Statistics with Vague Data. Reidel Publishing Company, Dordrecht 1987.

15. Kruse, R.; Gebhardt, J.; Klawonn, F. : Fuzzy-Systeme. B.G. Teubner Verlag, Stuttgart 1993

16. Malinvaud, E. : Statistical Methods of Econometrics. North-Holland, Amsterdam 1966

17. Nagel, M.; Feiler, D.; Bandemer, H.: Mathematische Statistik in der Umwelttechnik. In: Bandemer, H. (Ed.): Mathematik in der Technik, Bergakademie Freiberg 1985, Heft 1, 61-66

18. Otto, M.; Bandemer, H. : Pattern recognition based on fuzzy observations for spectroscopic quality control and chromographic fingerprinting. Ana. Chim. Acta 184 (1986), 21-31

19. Otto, M.; Bandemer, H. : A fuzzy approach to predicting chemical data from incomplete, uncertain and verbal compound features. In: Jochum, C.; Hicks, M.G.; Sunkel J. (Eds.) : Physical Property Prediction in Organic Chemistry, Springer Verlag, Berlin, Heidelberg 1988, 171-189

20. Serra, J. : Image Analysis and Mathematical Morphology; Vol. 2, Academic Press, Princeton, 1988.

21. Viertl, R. : Statistical inference for fuzzy data in environmetrics. Environmetrics 1 (1990), 37-42

22. Viertl, R. : Einführung in die Stochastik. Springer Verlag Wien, New York 1990

23. Viertl, R. : On statistical inference based on non-precise data. In: Bandemer. H. (Ed.): Modelling Uncertain Data, Akademie Verlag, Berlin 1993, 121-130

24. Wang, Z.; Klir, J. : Fuzzy Measure Theory, Plenum Press, New York 1992.

SOME NOTES ON POSSIBILISTIC LEARNING

J. Gebhardt and R. Kruse
University of Braunschweig, Braunschweig, Germany

Abstract

We outline a *possibilistic* learning method for structure identification from a database of samples. In comparison to the construction of Bayesian belief networks, the proposed framework has some advantages, namely the explicit consideration of *imprecise data*, and the realization of a controlled form of *information compression* in order to increase the efficiency of the learning strategy as well as approximate reasoning using local propagation techniques.

Our learning method has been applied to reconstruct a non–singly connected network of 22 nodes and 22 arcs without the need of any a priori supplied node ordering.

1 Introduction

Bayesian networks provide a well–founded normative framework for knowledge representation and reasoning with *uncertain*, but *precise* data. Extending pure probabilistic settings to the treatment of *imprecise* information usually restricts the computational tractability of the corresponding inference mechanisms. It is therefore near by hand to consider alternative uncertainty calculi that provide a justified form of *information compression* in order to support efficient reasoning in the presence of imprecise

This work has been partially funded by CEC–ESPRIT III Basic Research Project 6156 (DRUMS II)

and uncertain data without affecting the expressive power and correctness of decision making.

Such a modelling approach is appropriate for systems that accept *approximate* instead of crisp reasoning due to a non–significant sensitivity concerning slight changes of information. Possibility theory [1, 2] seems to be a promising framework for this purpose.

In this paper we focus our interest on the concept of a *possibilistic constraint network*, which is a dependency hypergraph and a family of possibility distributions that refer to the single hyperedges. Since the covering of all aspects of possibilistic modelling is beyond the scope of this paper, we will confine to the problem of learning the structure of a possibilistic network from data.

In section 2 we introduce a possibilistic interpretation of databases of imprecise samples. Based on this semantic background, section 3 briefly addresses the learning method. We mention some basic ideas and important results, which are presented in section 4, including an example of the successful application of our approach.

2 Possibilistic Interpretation of Sample Databases

Let $\text{Obj}(X_1, \ldots, X_n)$ be an *object type* of interest, which is characterized by a set $V = \{X_1, \ldots, X_n\}$ of *variables (attributes)* with finite domains $\Omega^{(i)} = \text{Dom}(X_i)$, $i = 1, \ldots, n$.

The *precise specification* of a current object state of this type is then formalized as a tuple $\omega_0 = (\omega_0^{(1)}, \ldots, \omega_0^{(n)})$, taken from the *universe of discourse* $\Omega = \Omega^{(1)} \times \ldots \times \Omega^{(n)}$. Any subset $R \subseteq \Omega$ can be used as a *set-valued specification* of ω_0, which consists of all states that are possible candidates for ω_0. R is therefore called *correct* for ω_0, if and only if $\omega_0 \in R$. R is called *imprecise*, iff $|R| > 1$, *precise*, iff $|R| = 1$, and *contradictory*, iff $|R| = 0$.

Suppose that general knowledge about dependencies among the variables is available in form of a database $\mathcal{D} = (D_j)_{j=1}^m$ of sample cases. Each case D_j is interpreted as a (set-valued) correct specification of a previously observed representative object state $\omega_j = (\omega_j^{(1)}, \ldots, \omega_j^{(n)})$.

Supporting imprecision (non-specificity) consists in stating $D_j = D_j^{(1)} \times \ldots \times D_j^{(n)}$, where $D_j^{(i)}$ denotes a nonempty subset of $\Omega^{(i)}$. We assume that $\omega_j^{(i)} \in D_j^{(i)}$ is satisfied, but no further information about any preferences among the elements in $D_j^{(i)}$ is given. When the cases in $\mathcal{D}$ are applied as an imperfect specification of the current object state ω_0, then *uncertainty* concerning ω_0 occurs in the way that the underlying frame conditions (here named as *contexts*, and denoted by c_j), in which the sample states ω_j have been observed, may only for some of the cases coincide with the context on which the observation of ω_0 is based on. The complete description of context c_j depends on the physical frame conditions of ω_j, but is also influenced by the frame conditions of observing ω_j by a human expert, a sensor, or any other observation unit. For the following consideration, we make some assumptions on the relationships between contexts and context-dependent specifications of object states. In particular, we suppose that our knowledge about ω_0 can be represented by an *imperfect specification*

$$
\begin{aligned}
\Gamma &= (\gamma, P_C), \\
C &= \{c_1, \ldots, c_m\}, \\
\gamma : C &\rightarrow \mathfrak{P}(\Omega), \\
\gamma(c_j) &= D_j, \qquad j = 1, \ldots, n,
\end{aligned}
$$

with C denoting the set of contexts, $\gamma(c_j)$ the context-dependent set-valued specification of ω_j, P_C a probability measure on C, and $\mathfrak{P}(\Omega)$ the power set of Ω. $P_C(\{c\})$ quantifies the probability of occurrence of context $c \in C$. If all contexts are in the same way representative and thus equally likely, then P_C should be the uniform distribution on C.

We suppose that C can be described as a subset of logical of propositions. The mapping $\gamma : C \rightarrow \mathfrak{P}(\Omega)$ indicates the assumption that there is a functional dependency of the sample cases from the underlying contexts, so that each context c_j uniquely determines its set-valued specification $\gamma(c_j) = D_j$ of ω_j. It is reasonable to state that $\gamma(c_j)$ is correct for ω_j (i.e.: $\omega_j \in \gamma(c_j)$) and of *maximum specificity*, which means that no proper subset of $\gamma(c_j)$ is guaranteed to be correct for ω_j with respect to context c_j. Related to the current object state of interest, specified by the (unknown) value $\omega_0 \in R$, and observed in a new context c_0, any c_j in C is adequate for delivering a set-valued specification of ω_0, if c_0 and c_j, formalized as logical propositions, are not contradicting. Intersecting the context-dependent set-valued specifications $\gamma(c_j)$ of all contexts c_j that do not contradict c_0, we obtain the most specific correct set-valued specification of ω_0 with respect to γ.

The idea of using set-valued mappings on probability fields in order to treat uncertain and imprecise data refers to similar random-set-like approaches that were suggested, for instance, in [3], [4], and [5]. But note that for operating on imperfect specifications in the field of knowledge-based systems, it is important to provide adequate semantics. We addressed this topic in more detail elsewhere [6, 7].

When we are only given a database $\mathcal{D} = (D_j)_{j=1}^m$ of sample cases, where $D_j \subseteq \Omega$ is assumed to be a context-dependent most specific specification of ω_j, we are normally not in the position to fully describe the contexts c_j in the form of propositions. For this reason it is convenient to carry out an information compression by paying attention to the context-dependent specifications rather than to the contexts themselves. We do not directly refer to $\Gamma = (\gamma, P_C)$, but to its degree of *α-correctness* w.r.t. ω_0, which is defined as the total mass of all contexts c_j that yield a correct context-dependent specification $\gamma(c_j)$ of ω_0. If we are given any $\omega \in \Omega$, then $\Gamma = (\gamma, P_C)$ is called *α-correct* w.r.t. ω, iff

$$
P_C(\{c \in C \mid \omega \in \gamma(c)\}) \geq \alpha, \qquad 0 \leq \alpha \leq 1.
$$

Note that, although the application of a probability measure P_C suggests disjoint contexts, we do not make any assumptions about the interrelation of contexts. With respect to the frame conditions that cause the observations, the contexts may be identical, partially corresponding, or disjoint. We add their weights, because disjoint contexts are the "worst case" in which we can not restrict the total weight to a smaller value without losing correctness. In this manner, a possibility degree is the upper bound for the total weight of the combined contexts.

Suppose that our only information about ω_0 is the α-correctness of Γ w.r.t. ω_0, without having any knowledge of the description of the contexts in C. Under these restrictions, we are searching for the most specific set-valued specification $A_\alpha \subseteq \Omega$ of ω_0, namely the largest subset of Ω such that α-correctness of Γ w.r.t. ω is satisfied for all $\omega \in A_\alpha$. It easily turns out that the family $(A_\alpha)_{\alpha \in [0,1]}$ consists of all α-cuts $[\pi_\Gamma]_\alpha$ of the induced *possibility distribution*

$$\pi_\Gamma : \Omega \rightarrow [0,1]$$
$$\pi_\Gamma(\omega) = P_C(\{c \in C \mid \omega \in \gamma(c)\}),$$

where for any π, taken from the set $\mathrm{POSS}(\Omega)$ of all possibility distributions that can be induced from imperfect specifications w.r.t. Ω, the α-cut $[\pi]_\alpha$ is defined as

$$[\pi]_\alpha = \{\omega \in \Omega \mid \pi(\omega) \geq \alpha\}, \qquad 0 < \alpha \leq 1,$$
$$[\pi]_0 = \Omega.$$

Note that $\pi_\Gamma(\omega)$ can in fact be viewed as a degree of possibility for the truth of "$\omega = \omega_0$": If $\pi_\Gamma(\omega) = 1$, then $\omega \in \gamma(c)$ holds for all contexts $c \in C$, which means that $\omega = \omega_j$ is possible for all sample object states ω_j, $j = 1, \dots, m$, so that $\omega_0 = \omega$ should be possible without any restriction.

If $\pi_\Gamma(\omega) = 0$, then $\omega_j = \omega$ has been rejected for ω_j, $j = 1, \dots, n$, since $\omega \notin \gamma(c_j)$ is true for the set-valued specifications $\gamma(c_j)$ of ω_j. This entails the impossibility of $\omega_0 = \omega$, if the description of the context c_0 for the specification of ω_0 is assumed to be a conjunction of the descriptions of any contexts in C.

If $0 < \pi_\Gamma(\omega) < 1$, then there are contexts that support $\omega_0 = \omega$ as well as contexts that contradict $\omega_0 = \omega$. The quantity $\pi_\Gamma(\omega)$ reflects the maximum possible total mass of contexts that support $\omega_0 = \omega$.

Ignoring the semantics of consideration contexts, note that π_Γ formally coincides with the *one-point coverage* of Γ [8], interpreted as a (non-nested) random set. There are also some connections to *contour functions* of consonant belief functions [9] and *falling shadows* in set valued statistics [10]. On the other hand, from a semantic point of view, operating on possibility distributions may better be oriented at the concept of α-correctness. For an extensive presentation of this view of possibility theory and maximum-specificity-preserving operations on possibility distributions based on α-correctness, we refer to [11, 12].

Using π_Γ as a compressed representation of the information contained in $\mathcal{D}$, we do not model existing dependencies among the considered attributes in an explicit way.

Suppose that such dependencies can be represented in terms of a hypergraph $H = (V, \mathcal{E})$, so that any most specific set-valued specification $R_0 \subseteq \Omega$ of ω_0 is assumed to have a *lossless join decomposition* $(R_E)_{E \in \mathcal{E}}$, which means that

$$R_0 = \{\omega \in \Omega \mid \forall E \in \mathcal{E} : \textstyle\prod_E(\omega) \in R_E\},$$

where $\prod_E(\omega)$ denotes the projection of ω onto $\Omega^E \overset{\text{Df}}{=} \underset{X_i \in E}{\times} \Omega^{(i)}$.

Given the pair $(\mathcal{D}, H)$, the database $\mathcal{D}$ may be viewed as a family of databases $\mathcal{D}_E$, $E \in \mathcal{E}$, with each $\mathcal{D}_E$ providing samples of observed relationships among the attributes contained in E. In this case we have imperfect specifications $\Gamma_E = (\gamma_E, P_E)$, $C_E = \{c_1^E, \ldots, c_m^E\}$, $\gamma_E : C \to \mathcal{P}(\Omega^E)$, $P_E(\{c_j^E\}) = \frac{1}{m}$, $\gamma_E(c_j^E) = \prod_E(D_j)$, $j = 1, \ldots, m$. The family $(\Gamma_E)_{E \in \mathcal{E}}$ induces a *possibilistic constraint network* $(\pi_{\Gamma_E})_{E \in \mathcal{E}}$ as the possibilistic interpretation of $\mathcal{D}$ given H.

In the following section we consider the problem of learning the dependency structure H of such a network from $\mathcal{D}$.

3   Inducing Possibilistic Networks from Data

Algorithms for the construction of *probabilistic* networks from data are based on linearity and normality assumptions [13], the extensive testing of conditional independence relations [14, 15], or Bayesian approaches [16, 17]. Some crucial problems of these methods refer to their complexity, their reliability unless the volume of data is enormous, and essential presuppositions like the requirement of an ordering of the nodes and a priori distribution assumptions. Some heuristic methods such as, for example, the K2 greedy algorithm [16] and the combination of conditional independence tests with Bayesian learning [18] have been provided in order to overcome complexity problems. Handling not only uncertain, but also imprecise information, the computational effort needed for learning possibilistic networks is presumably similar. Since the quite restrictive case of decomposing relational data into a structure taken from a class of schemes has turned out to be intractable, even if each member of the class is tractable [19] — tree structured schemes are an exception —, exact structure identification of possibilistic networks is insufficient in the general case. Considering the approximate reasoning property of possibilistic inference, the problem of exact structure identification is weakened in the sense that wrong structures, obtained by ignoring relevant dependencies or adding irrelevant dependencies, always yield correct, but perhaps less specific and therefore non–informative inference results.

Our idea of inducing an appropriate network structure is to determine a *directed acyclic graph* (DAG) $G = (V, E)$ that best fits existing causal dependencies in the database $\mathcal{D}$, and then to transform G into a hypergraph $H = (V, \mathcal{E})$, so that the pair $(\mathcal{D}, H)$ induces an optimal possibilistic network $(\pi_{\Gamma_E})_{E \in \mathcal{E}}$. G is regarded as being optimal, if it minimizes the expected amount of additional information that is necessary in order to identify any element $\omega_0 \in \Omega$, when passing G from the root nodes to the leafs, considering the constraints that are available via the attached (conditioned) possibility distributions. This expected amount of information Nonspec$(G|\mathcal{D})$ is quantified in a natural way with the aid of *Hartley information* [20] as an appropriate measure of nonspecifity for imprecise data. We will first define it for a structure $(G, \mathcal{R})$, where $\mathcal{R}$ is a family of relations, before extending it to the possibilistic setting.

Definition
Let $G = (V, E)$ be a directed acyclic graph. For each $X_i \in V$ let par(X_i) denote the set of parents of X_i. Furthermore let $\mathcal{R} = (R_i)_{i=1}^n$ be a family of non–empty relations $R_i \subseteq \Omega^{W_i}$, where $W_i = \text{par}(X_i) \cup \{X_i\}$. Then, given $\omega_0 \in \Omega$,

$$\text{Nonspec}(\omega_0 \mid G, \mathcal{R}) \stackrel{\text{Df}}{=} \sum_{i=1}^n H(\omega_0 \mid G, R_i),$$

$$H(\omega_0 \mid G, R_i) \stackrel{\text{Df}}{=} \log_2 \left(\left| \Pi_{\{X_i\}} \left\{ \omega \in R_i \mid \Pi_{\text{par}(X_i)}(\omega_0) = \Pi_{\text{par}(X_i)}(\omega) \right\} \right| \right)$$

is called the *degree of nonspecifity of* ω_0 *w.r.t. G and $\mathcal{R}$.*

For generalizing Nonspec$(\omega_0|G, \mathcal{R})$ to a family of possibility distributions, we consider our possibilistic interpretation of $\mathcal{D}$ introduced in the previous section. If we are given a DAG G, then we suppose that there are underlying imperfect specifications $\Gamma_i = (\gamma_i, P_i)$ of sample object states $\omega_1, \ldots, \omega_m$, defined by $\gamma_i : C_i \rightarrow \mathcal{P}(\Omega^{W_i})$, $\gamma_i(c_j^{(i)}) = \Pi_{W_i}(D_j)$, $P_i(\{c_j^{(i)}\}) = \frac{1}{m}$, $j = 1, \ldots, m$, $i = 1, \ldots, n$, with respect to sets $C_i = \{c_1^{(i)}, \ldots, c_m^{(i)}\}$ of equally likely consideration contexts.
It is convenient to assume that we are a priori indifferent concerning the choice of a family $(\alpha_i)_{i=1}^n$ of possible correctness degrees of Γ_i w.r.t. $\Pi_{W_i}(\omega_0)$. Nonspec$(\omega_0|G, \mathcal{R})$ can be calculated for all families $\mathcal{R} = ([\pi_{\Gamma_i}]_{\alpha_i})_{i=1}^n$ of relations induced by the different families $(\alpha_i)_{i=1}^n$.
Furthermore it is reasonable to assume that we do not have any preferences regarding the current object state ω_0 in our a priori state of knowledge. Stating therefore the uniform distribution assumptions on Ω and the families of possible correctness degrees, the task is to compute the resulting expected value of nonspecifity Nonspec$(G|\mathcal{D})$ that evaluates network structure G, given the database $\mathcal{D}$.

4   Results and Concluding Remarks

- Let $\mathcal{G}_k(V)$, $k \geq 2$, denote the class of all directed acyclic graphs w.r.t. V that satisfy the condition $|\text{par}(X)| \leq k-1$ for all $X \in V$. We developed an *Algorithm G1* for determining a DAG $G \in \mathcal{G}_k(V)$ that minimizes $E(\text{Nonspec}[\mathcal{N}_G(D)])$ among all DAGs in $\mathcal{G}_k(V)$ that satisfy the node ordering constraints of G. Algorithm G1 has a time complexity of $O(mn^k r^k)$, where r is the maximum cardinality of the involved domains. It does not need any presupposed node ordering, and although it is only optimal w.r.t. a subclass of $\mathcal{G}_k(V)$, it nevertheless tends to deliver a good choice w.r.t. $\mathcal{G}_k(V)$.

- Elementing arcs of a DAG $G \in \mathcal{G}_k(V)$ in order to get a more simple DAG $G' \in \mathcal{G}_{k-1}(V)$, is connected with a loss of information which is quantified by the corresponding increasement of $E(\text{Nonspec}[\mathcal{N}_{G'}(D)])$ compared to $E(\text{Nonspec}[\mathcal{N}_G(D)])$.

- Algorithm G1 can be made more efficient with the aid of a Greedy search method, starting with $\mathcal{G}_2(V)$ and stepwise extending the optimal output graph with respect to those arcs that reflect the strongest causal dependencies (i.e. the smallest degree of nonspecificity) to the class $\mathcal{G}_k(V)$. The resulting *Algorithm G2* has a time complexity of $O(mn^2 r^k)$.

- From a graph theoretical point of view, independence in the DAG of a possibilistic causal network is represented in the same way as independence in Bayesian networks. Due to the different uncertainty calculus, independence in our approach turns out to basically coincide with the concept of *non-interactivity* well-known from possibility theory. Non-interactivity satisfies the basic properties of independence as proposed in [21], with the exception of the intersection axiom.

- From database theory it is well-known that given a relation R and any hypergraph H, deciding whether rel$(\mathcal{N}_H(R)) = R$ is NP-hard. Constructing a lossless join decomposition of a relation within a class of dependency hypergraphs is presumably intractable even in cases where each individual member of the class is tractable [19].

  Since the problem of structure identification in relational data can be viewed as a special case of the corresponding problem in the possibilistic framework, it is out of reach to get a general, non-heuristic, efficient algorithm for possibilistic structure identification. One of the tasks of our future research in this field therefore consists in working out, how far Algorithms G1 and G2 deliver tight approximations of optimal (i.e. maximum specificity preserving) decompositions.

- Algorithms G1 and G2 have successfully been applied for reconstructing the network shown in Figure 1. The underlying application refers to a Bayesian approach, implemented with HUGIN [22] for daily use in Denmark. It deals with the determination of the genotype and verifying the parentage in the F-blood group system of Danish Jersey Cattle [23]. The application is supported by a real database of 747 sample cases for the 9 attributes marked as grey nodes in Figure 1, including lots of missing values. There is also additional expert knowledge regarding the quantitative dependencies among other attributes. Using this information we extended the database to an artificial database for all attributes. Running Algorithm G2 on this database, the network could be efficiently reconstructed in the possibilistic setting without erraneous links, except from those dependencies, where a unique directing of arcs is not possible, since not expressable in a database. We will include our learning strategy in the system POSSINFER [24], a software tool for *possibilistic inference* that we develop in cooperation with Deutsche Aerospace in the field of data fusion problems.

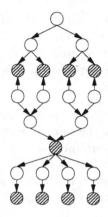

Fig. 1,   *The blood–type example network*

Acknowledgements

We would like to thank S.L. Lauritzen and L.K. Rasmussen for supporting us with respect to the bloodtype determination example.

References

[1]   Zadeh, L.A.: Fuzzy Sets as a Basis for a Theory of Possibility, Fuzzy Sets and Systems, 1 (1978), 3–28.

[2]   Dubois, D. and H. Prade: Possibility Theory, Plenum Press, New York 1988.

[3]   Strassen, V.: Meßfehler und Information, Zeitschrift Wahrscheinlichkeitstheorie und verwandte Gebiete, 2 (1964), 273–305.

[4]   Dempster, A.P.: Upper and Lower Probabilities Induced by a Random Closed Interval, Ann. Math. Stat., 39 (1968), 957–966.

[5]   Kampé de Fériet, J.: Interpretation of Membership Functions of Fuzzy Sets in Terms of Plausibility and Belief, in: Fuzzy Information and Decision Processes (Ed. M.M. Gupta, E. Sanchez), North–Holland, 1982, 13–98.

[6]   Gebhardt, J. and R. Kruse: The Context Model – An Integrating View of Vagueness and Uncertainty, Int. J. of Approximate Reasoning, 9 (1993), 283–314.

[7]   Gebhardt, J. and R. Kruse: A Comparative Discussion of Combination Rules in Numerical Settings, CEC–ESPRIT III BRA 6156 DRUMS II, Annual Report, 1993.

[8]   Nguyen, H.T.: On Random Sets and Belief Functions, J. of Mathematical Analysis and Applications, 65 (1978), 531–542.

[9]   Shafer, G.: A Mathematical Theory of Evidence, Princeton University Press, Princeton 1976.

[10] Wang, P.Z.: From the Fuzzy Statistics to the Falling Random Subsets, in: Advances in Fuzzy Sets, Possibility and Applications (Ed. P.P. Wang), Plenum Press, New York 1983, 81–96.

[11] Gebhardt, J. and R. Kruse: A New Approach to Semantic Aspects of Possibilistic Reasoning, in: Symbolic and Quantitative Approaches to Reasoning and Uncertainty (Ed. M. Clarke, R. Kruse, and S. Moral), Lecture Notes in Computer Science, 747 (1993), Springer, Berlin, 151–160.

[12] Gebhardt, J. and R. Kruse: On an Information Compression View of Possibility Theory, in: Proc. 3rd IEEE Int. Conf. on Fuzzy Systems, Orlando 1994.

[13] Pearl, J. and N. Wermuth: When Can Association Graphs Admit a Causal Interpretation (First Report), in: Preliminary Papers of the 4th Int. Workshop on Artificial Intelligence and Statistics, Ft. Lauderdale, FL January 3–6 1993, 141–151.

[14] Verma, T. and J. Pearl: An Algorithm for Deciding if a Set of Observed Independencies Has a Causal Explanation, in: Proceedings 8th Conf. on Uncertainty in AI, 1992, 323–330.

[15] Spirtes, P. and C. Glymour: An Algorithm for Fast Recovery of Sparse Causal Graphs, Social Science Computing Review, 9 (1991), 62–72.

[16] Cooper, C. and E. Herskovits: A Bayesian Method for the Induction of Probabilistic Networks from Data, Machine Learning, 9 (1992), 309–347.

[17] Lauritzen, S.L., B. Thiesson, and D. Spiegelhalter: Diagnostic Systems Created by Model Selection Methods — A Case Study, in: Preliminary Papers of the 4th Int. Workshop on Artificial Intelligence and Statistics, Ft. Lauderdale, FL January 3–6 1993, 141–151.

[18] Singh, M. and M. Valtorta: An Algorithm for the Construction of Bayesian Network Structures from Data, in: Proc. 9th. Conf. on Uncertainty in Artificial Intelligence, Washington 1993, 259–265.

[19] Dechter, R. and J. Pearl: Structure Identification in Relational Data, Artificial Intelligence, 58 (1992), 237–270.

[20] Hartley, R.V.L., Transmission of Information: The Bell Systems Technical J., 7 1928, 535–563.

[21] Pearl, J.: Probabilistic Reasoning in Intelligent Systems — Networks of Plausible Inference, Morgan Kaufman, San Mateo 1988.

[22] Andersen, S.K., K.G. Olesen, F.V. Jensen, and F. Jensen: HUGIN — A Shell for Building Bayesian Belief Universes for Expert Systems, in: Proc. 11th Int. Joint Conf. on AI, 1989, 1080–1085.

[23] Rasmussen, L.K.: Blood Group Determination of Danish Jersey Cattle in the F-blood Group System, Dina Research Report 8, Dina Foulum, 8830 Tjele, Denmark, November 1992.

[24] Kruse, R., J. Gebhardt, and F. Klawonn: Foundations of Fuzzy Systems, Wiley, Chichester 1994.

STATISTICS WITH FUZZY DATA

R. Viertl
Technical University of Wien, Wien, Austria

Abstract

Besides stochastic variation of real data there is another kind of uncertainty in observations called *imprecision* and the corresponding measurement results are called *non-precise data*. Usually this kind of uncertainty is not described in statistics. Especially in environment statistics, where also very small quantities are measured, this imprecision cannot be neglected. Otherwise by extrapolations the precision of results is very misleading. Therefore statistical methods have to be generalized for non-precise data. The mathematical concept to describe non-precise data are so-called *non-precise numbers* and *non-precise vectors*. Situations are explained, where the characterizing functions of non-precise numbers can be given explicitly. Using the mathematical concept of non-precise numbers and vectors statistical procedures can be generalized to non-precise data. These generalizations are explained in the contribution.

1 Non-precise data

The results of measurements are often not precise numbers or vectors but more or less *non-precise*. This uncertainty is different from measurement errors and stochastic uncertainty and is called *imprecision*. Imprecision is a feature of single observations. Errors are described by statistical models and should not be confused with imprecision. In general imprecision and errors are superimposed. *In this paper errors are not considered.*

Example: The results of many measurement procedures are gray tone pictures (for example X-ray measurements). The resulting data are not real numbers but non-precise numbers.

Therefore it is necessary to model *non-precise numbers*.

A special case of non-precise data are *interval data*.

Precise real numbers $x_0 \in \mathbb{R}$ as well as intervals $[a, b] \subseteq \mathbb{R}$ are uniquely characterized by their *indicator functions* $I_{\{x_0\}}(\cdot)$ and $I_{[a,b]}(\cdot)$ respectively. The indicator function $I_A(\cdot)$ of a classical set A is defined by

$$I_A(x) = \begin{cases} 1 & \text{for } x \in A \\ 0 & \text{for } x \notin A. \end{cases}$$

The imprecision of measurements implies that exact boundaries of interval data are not realistic. Therefore it is necessary to generalize real numbers and intervals to describe imprecision. This is done by the concept of *non-precise numbers* and general *non-precise subsets* of $\mathbb{R}$ as generalizations of intervals. Such non-precise subsets are also called *fuzzy sets*. Non-precise numbers as well as non-precise subsets of $\mathbb{R}$ are described by generalizations of indicator functions, called *characterizing functions*.

2   Non-precise numbers and characterizing functions

To model non-precise observations so-called *non-precise numbers* x^* are quantified by *characterizing functions* $\xi(\cdot)$, which characterize the imprecision of an observation.

Definition 1: A *characterizing function* $\xi(\cdot)$ of a non-precise number is a real function of a real variable with the following properties:

(1) $\xi : \mathbb{R} \to [0, 1]$

(2) $\exists\, x_0 \in \mathbb{R} : \xi(x_0) = 1$

(3) $\forall\, \alpha \in (0, 1]$ the set $B_\alpha := \{x \in \mathbb{R} : \xi(x) \geq \alpha\} = [a_\alpha, b_\alpha]$ is a closed interval, called α-*cut* of $\xi(\cdot)$.

The set $supp\big(\xi(\cdot)\big) := \{x \in \mathbb{R} : \xi(x) > 0\}$ is called *support* of $\xi(\cdot)$.

Remark: Non-precise observations and non-precise numbers will be marked by stars, i.e. x^*, to distinguish them from (precise) real numbers x. Every non-precise number x^* is characterized by the corresponding characterizing function $\xi_{x^*}(\cdot)$.

Remark: Non-precise numbers are special fuzzy subsets of $\mathbb{R}$. This is reasonable as generalization of intervals, because intervals are classical subsets of $\mathbb{R}$.

Proposition 1: Characterizing functions $\xi(\cdot)$ are uniquely determined by the family $\big(B_\alpha;\ \alpha \in (0, 1]\big)$ of their α-cuts B_α and the following holds

$$\xi(x) = \max_{\alpha \in (0,1]} \alpha \cdot I_{B_\alpha}(x) \qquad \forall\ x \in \mathbb{R}.$$

Proof: Let $x_0 \in \mathbb{R}$ then it follows

$$\alpha \cdot I_{B_\alpha}(x_0) = \alpha \cdot I_{\{x : \xi(x) \geq \alpha\}}(x_0) = \begin{cases} \alpha & \text{for} \quad \xi(x_0) \geq \alpha \\ 0 & \text{for} \quad \xi(x_0) < \alpha \end{cases}$$

and from that $\alpha \cdot I_{B_\alpha}(x_0) \leq \xi(x_0) \quad \forall \; \alpha \in (0,1]$ and $\sup_{\alpha \in (0,1]} \alpha \cdot I_{B_\alpha}(x_0) \leq \xi(x_0)$.

For $\alpha_0 = \xi(x_0)$ we obtain $B_{\alpha_0} = \{x : \xi(x) \geq \xi(x_0)\} = [a_{\alpha_0}, b_{\alpha_0}]$ and therefore $\alpha_0 \cdot I_{B_{\alpha_0}}(x_0) = \alpha_0 \cdot 1 = \xi(x_0) = \max_{\alpha \in (0,1]} \alpha \cdot I_{B_\alpha}(x_0)$.

2.1 Special non-precise numbers

Non-precise numbers x^* can be described by corresponding characterizing functions $\xi_{x^*}(\cdot)$ in the following form

$$\xi_{x^*}(x) = \begin{cases} L(x) & \text{for} \quad x \leq m_1 \\ 1 & \text{for} \quad m_1 \leq x \leq m_2, \quad \text{with } m_1 \leq m_2 \\ R(x) & \text{for} \quad x \geq m_2 \end{cases}$$

where $L(\cdot)$ is an increasing function and $R(\cdot)$ a decreasing real function.

The following special forms of characterizing functions are frequently used:

Trapezoid non-precise numbers $t^*(m_1, m_2, a_1, a_2)$

$$L(x) = \max\left(0, \frac{x - m_1 + a_1}{a_1}\right) \qquad \text{for} \qquad x \leq m_1$$

$$R(x) = \max\left(0, \frac{m_2 + a_2 - x}{a_2}\right) \qquad \text{for} \qquad x \geq m_2$$

For $m_1 = m_2$ so called *triangular non-precise numbers* $t^*(m, a_1, a_2)$ are obtained.

Exponential non-precise numbers $e^*(m_1, m_2, a_1, a_2, p_1, p_2)$

$$L(x) = exp\left(-|\frac{x - m_1}{a_1}|^{p_1}\right)$$

$$R(x) = exp\left(-|\frac{x - m_2}{a_2}|^{p_2}\right)$$

The problem of how to obtain the characterizing function of a non-precise observation is discussed in section 3.

2.2 Convex hull of a fuzzy number

For functions $\varphi(\cdot)$ which fulfill conditions (1) and (2) of definition 1 but not condition (3) in the following way $\varphi(\cdot)$ can be transformed into a characterizing function $\xi(\cdot)$ fulfilling also condition (3).

Definition 2: Let $\varphi(\cdot)$ be a real function fulfilling conditions (1) and (2) of definition 1 but not condition (3). Then some α-cuts B_α of $\varphi(\cdot)$ are unions of disjoint intervals $B_{\alpha,i} = [a_{\alpha,i}; b_{\alpha,i}]$, i.e. $B_\alpha = \bigcup\limits_{i=1}^{k_\alpha} B_{\alpha,i}$. Using proposition 1 the so called *convex hull* $\xi(\cdot)$ of $\varphi(\cdot)$ is defined via the α-cuts $C_\alpha(\xi(\cdot))$ by

$$C_\alpha := \left[\min_{i=1(1)k_\alpha} a_{\alpha,i} \ ; \ \max_{i=1(1)k_\alpha} b_{\alpha,i} \right].$$

Therefore $\xi(\cdot)$ is given by $\xi(x) = \max\limits_{\alpha \in (0,1]} \alpha \cdot I_{C_\alpha}(x) \qquad \forall \ \ x \in \mathbb{R}$.

3 Construction of characterizing functions

The construction of characterizing functions to observed non-precise data is depending on the field of application. Looking to real examples one can see some methodology.

Example: For the non-precise quantity water level of a river one can look at the intensity of wetness of the survey rod. The wetness $w(h)$ as a function of the height h can be used to obtain the characterizing function of the quantity water level in the following way: Take the derivative of the function $w(\cdot)$. Normalizing this function by dividing it by its maximal value we obtain a characterizing function which describes the non-precise observation water level.

Example: Biological life times are important examples of non-precise data. The observation of the life time of a tree is connected with this kind of uncertainty. Often the end of a life time of a system is characterized by the degradation of a certain quantity which is measured continuously.

Remark: For functions $\xi(\cdot)$ obtained as in the examples or similar results which do not fulfill condition (3) of definition 1 the *convex hull* of $\xi(\cdot)$ can be used. This convex hull is a non-precise number in the sense of definition 1.

4 Non-precise vectors and functions

4.1 Non-precise vectors

Many statistical data are – in case of idealized precise observations – given by vectors of real numbers. Real multi-dimensional measurement data are mostly non-precise. Therefore this imprecision has to be described.

Example: Measuring the location of a point in the plane one obtains idealized a pair $(x, y) \in \mathbb{R}^2$. For real observations one obtains – for example by a radar equipment – a light point with finite range. This is a non-precise 2-dimensional vector $\underline{x}^\star$ with *characterizing function* $\xi_{\underline{x}^\star}(x_1, x_2)$ which describes the imprecision of the observation.

For the mathematical description of vector-valued non-precise observations one can use so-called *non-precise vectors* which are described by *characterizing functions* of non-precise vectors.

Definition 3: A non-precise n-dimensional vector $\underline{x}^\star$ is determined by the corresponding *characterizing function* $\xi_{\underline{x}^\star}(\cdot)$ of n real variables with the following properties

(1) $\xi_{\underline{x}^\star} : \mathbb{R}^n \to [0, 1]$

(2) $\exists\, \underline{x}_0 \in \mathbb{R}^n : \xi_{\underline{x}^\star}(x_0) = 1$

(3) $B_\alpha(\underline{x}^\star) := \{\underline{x} \in \mathbb{R}^n : \xi_{\underline{x}^\star}(\underline{x}) \geq \alpha\}$ is a simply connected and compact subset
$$\text{of } \mathbb{R}^n \qquad \forall\, \alpha \in (0, 1]\,.$$

The set $supp\,(\xi_{\underline{x}^\star}(\cdot)) := \{\underline{x} \in \mathbb{R}^n : \xi_{\underline{x}^\star}(\underline{x}) > 0\}$ is called *support* of $\xi_{\underline{x}^\star}(\cdot)$.

Proposition 2: Characterizing functions $\xi_{\underline{x}^\star}(\cdot)$ of non-precise vectors $\underline{x}^\star$ are also determined by their families $(B_\alpha(\underline{x}^\star);\quad \alpha \in (0, 1])$ of α-cuts and the following is valid

$$\xi_{\underline{x}^\star}(\underline{x}) = \max_{\alpha \in (0,1]} \alpha \cdot I_{B_\alpha(\underline{x}^\star)}(\underline{x}) \qquad \forall\ \underline{x} \in \mathbb{R}^n.$$

The proof is similar to the proof of proposition 1.

4.2 Functions of non-precise variables

In statistics often functions of observations are important. An example is the sample mean $\bar{x}_n$ of n observations $x_1, \cdots, x_n$ of a stochastic quantity.

In case of non-precise observations $x_1^\star, \cdots, x_n^\star$ we have a function of n non-precise quantities and the value of the function is a non-precise element of the domain. This kind of generation of imprecision of functions is the one, induced by the imprecision of argument values of a function.

Definition 4: Let $g : \mathbb{R}^n \to \mathbb{R}$ be a classical function whose argument values $\underline{x}$ are not known precisely but are non-precise vectors $\underline{x}^\star$ in $\mathbb{R}^n$. For non-precise argument value $\underline{x}^*$ with corresponding characterizing function $\xi(\cdot)$ the non-precise value $g(\underline{x}^*)$ of the function is given by the characterizing function $\psi(\cdot)$ whose values are

$$\psi(y) := \sup\{\xi(\underline{x}): \ \underline{x} \in \mathbb{R}^n \wedge \ g(\underline{x}) = y\}.$$

Remark: For general functions $g(\cdot)$ it is not provided that $\psi(\cdot)$ is a characterizing function in the sense of definition 1. For *continuous functions* the conditions are fulfilled. This is shown in the following proposition.

Proposition 3: Let $\underline{x}^*$ be a non-precise vector in $\mathbb{R}^n$ with characterizing function $\xi(\cdot)$ and $g : M \rightarrow \mathbb{R}$ a continuous function with $M \subseteq \mathbb{R}^n$ and $supp(\xi(\cdot)) \subseteq M$. Then the function $\psi(\cdot)$ from definition 4 is a characterizing function in the sense of definition 1 which defines a non-precise number y^*. For the α-cut-presentation

$$\left(B_\alpha(y^*); \ \alpha \in (0,1]\right)$$

of y^* the following intervals are obtained:

$$B_\alpha(y^*) = \left[\min_{\underline{x} \in B_\alpha(\underline{x}^*)} g(\underline{x}) \ , \ \max_{\underline{x} \in B_\alpha(\underline{x}^*)} g(\underline{x})\right].$$

Proof: The conditions (1) to (3) of characterizing functions from definition 1 have to be shown. Condition (1) and condition (2) are trivially fulfilled. By the continuity of $g(\cdot)$ the set $g^{-1}(\{y\})$ is closed and therefore we obtain $\sup\{\xi(\underline{x}): \underline{x} \in g^{-1}(\{y\})\} = \max\{\xi(\underline{x}): \underline{x} \in g^{-1}(\{y\})\}$.
Next we show $B_\alpha(y^*) = g\left(B_\alpha(\underline{x}^*)\right) \ \forall \ \alpha \in (0,1]$.

For $\alpha \in (0,1]$ and $y \in g\left(B_\alpha(\underline{x}^*)\right)$ there exists an $\underline{x} \in B_\alpha(\underline{x}^*)$ with $y = g(\underline{x})$. From that follows $\xi(\underline{x}) \geq \alpha$ and $\sup\{\xi(\underline{x}): g(\underline{x}) = y\} \geq \alpha$, and by definition 4 $\psi(y) \geq \alpha$ which yields $y \in B_\alpha(y^*) \Rightarrow g\left(B_\alpha(\underline{x}^*)\right) \subseteq B_\alpha(y^*)$.

On the other side for $y \in B_\alpha(y^*)$ we have $\psi(y) \geq \alpha$ and $\sup\{\xi(\underline{x}): g(\underline{x}) = y\} \geq \alpha$. By the continuity of $g(\cdot)$ it follows $\max\{\xi(\underline{x}): g(x) = y\} = \sup\{\xi(\underline{x}): g(\underline{x}) = a\} \geq \alpha$. Therefore there exists an $\underline{x}_0 : \xi(\underline{x}_0) \geq \alpha$ with $g(\underline{x}_0) = y$ and $\underline{x}_0 \in B_\alpha(\underline{x}^*)$ and $y \in g\left(B_\alpha(\underline{x}^*)\right) \Rightarrow B_\alpha(y^*) \subseteq g\left(B_\alpha(\underline{x}^*)\right)$.

By the continuity of $g(\cdot)$ and the fact that $B_\alpha(\underline{x}^*)$ is a compact and simply connected subset of $\mathbb{R}^n$ it follows that $g\left(B_\alpha(\underline{x}^*)\right)$ is a compact and simply connected subset of $\mathbb{R}$ and therefore a closed interval of the form given in the proposition.

Remark: In general it can be complicated to obtain the analytical form of $\psi(\cdot)$ from definition 4. For linear functions

$$g(\underline{x}) = A\underline{x} = \sum_{i=1}^{n} a_i x_i \ \text{ with } \ A = (a_1, \cdots, a_n), \ a_i \in \mathbb{R},$$

this is simple.

For general functions the imprecision of the result can be graphically represented by a finite number of α-cuts of $\psi(\cdot)$.

4.3   Non-precise functions

More generally non-precise functions can be described as generalizations of classical functions $g : M \to \mathbb{R}$ in the following way:

Definition 5: A *non-precise function* $g^*(\cdot)$ is a mapping which assigns to every element $x \in M$ a non-precise number $g^*(x)$.

Remark: Non-precise functions are uniquely given by the family

$$\left(\phi_x(\cdot); \ x \in M \right)$$

of non-precise numbers y_x^* with corresponding characterizing functions $\phi_x(\cdot)$. For graphical presentations of non-precise functions the description of so-called α-*level-curves* is of advantage. These α-level-curves are defined as follows:
For $\alpha \in (0,1]$ consider the α-cuts

$$B_\alpha(y^\star) = \left[\ \underline{g}_\alpha(x), \ \overline{g}_\alpha(x) \ \right] \quad \text{for every} \ \ x \in M.$$

Then for variable x two classical functions $\underline{g}_\alpha(x)$ and $\overline{g}_\alpha(x)$ are obtained.

The graphs of these functions are called α-*level-curves* of the non-precise function $g^*(\cdot)$.

5   Combination of non-precise observations

Observations of a one-dimensional stochastic quantity (also called random quantity) are often non-precise. Such data are described by non-precise numbers x^* and n observations by n non-precise numbers $x_1^\star, \cdots, x_n^\star$ with corresponding characterizing functions $\xi_1(\cdot), \cdots, \xi_n(\cdot)$. These n observations are also called *non-precise sample* or *non-precise data* and are denoted by $D^\star$.

Remark: There can be also precise observations $x_i \in \mathbb{R}$ in the sample. In this case the corresponding characterizing function is the one-point indicator function $I_{\{x_i\}}(\cdot)$. The characterizing functions of the non-precise numbers can have intersecting supports.

In statistical inference functions $t(x_1, \cdots, x_n)$ of observations $x_1, \cdots, x_n$, so-called *statistics*, are important.

For precise observations $x_1, \cdots, x_n$ with $x_i \in M$, where M is the *observation space*, statistics $t(x_1, \cdots, x_n)$ are measurable functions from the *sample space* $M^n = M \times \cdots \times M$ into another measurable space, for example $\mathbb{R}$.

Examples for $t(x_1, \cdots, x_n)$ are the *sample mean* $\overline{x}_n$ defined by

$$\overline{x}_n = t_1(x_1, \cdots, x_n) := \frac{1}{n} \sum_{i=1}^{n} x_i$$

and the *sample variance* s_n^2 defined by

$$s_n^2 = t_2(x_1, \cdots, x_n) := \frac{1}{n-1} \sum_{i=1}^{n} (x_i - \bar{x}_n)^2.$$

For non-precise observations it is necessary to generalize functions. In order to do that definition 4 from section 4.2 can be used.

The generalization of statistical functions to the situation of non-precise data makes it necessary to combine the n non-precise elements of the observation space M to a non-precise element of the sample space M^n in order to allow the application of definition 4 and proposition 3.

Combining n non-precise observations $x_1^\star, \cdots, x_n^\star$ with corresponding characterizing functions $\xi_1(\cdot), \cdots, \xi_n(\cdot)$ should be done reasonably in such a way that the resulting characterizing function $\xi : M^n \to [0,1]$ of the *non-precise combined element* $\underline{x}^\star$ of the sample space M^n is a non-precise vector in the sense of definition 3 from section 4.1.

In general reasonable *combination-rules* $K_n(\cdot, \cdots, \cdot)$ in

$$\xi(x_1, \cdots, x_n) := K_n\big(\xi_1(x_1), \cdots, \xi_n(x_n)\big)$$

have to fulfill the following properties:

(1) $K_1\big(\xi_1(x)\big) = \xi_1(x)$

(2) $K_n\left(I_{\{\mathring{x}_1\}}(x_1), \cdots, I_{\{\mathring{x}_n\}}(x_n)\right) = I_{\{(\mathring{x}_1, \cdots, \mathring{x}_n)\}}(x_1, \cdots, x_n)$

(3) $K_n\left(I_{[a_1,b_1]}(x_1), \cdots, I_{[a_n,b_n]}(x_n)\right) = I_{[a_1,b_1] \times \cdots \times [a_n,b_n]}(x_1, \cdots, x_n).$

The following combination rules are used:

Product combination-rule: (short: *Product-rule*)

$$\xi(x_1, \cdots, x_n) := \prod_{i=1}^{n} \xi_i(x_i)$$

Minimum combination-rule (short: *Minimum-rule*)

$$\xi(x_1, \cdots, x_n) := \min_{i=1(1)n} \xi_i(x_i)$$

Remark: The advantage of the product-rule is a kind of consistency of estimators for increasing sample size n.

The minimum-rule is motivated by the fact that the α-cuts of the non-precise combined sample element are easy to obtain from the α-cuts of the non-precise observations. Moreover computations are easier using the minimum-rule.

Lemma 1: Let $x_1^\star, \cdots, x_n^\star$ be a sample of non-precise observations with characterizing functions $\xi_1(\cdot), \cdots, \xi_n(\cdot)$. If the minimum-rule is used to obtain the non-precise combined sample element $\underline{x}^\star$ then the α-cuts of $\underline{x}^\star$ are the Cartesian products of the α-cuts of the non-precise observations $x_i^\star$, i.e.

$$B_\alpha(\underline{x}^\star) = \mathrm{X}_{i=1}^n B_\alpha(x_i^\star) \qquad \forall \quad \alpha \in (0,1].$$

Proof: For all $\alpha \in (0,1]$ we have

$$\begin{aligned}
B_\alpha(\underline{x}^\star) &= \{\underline{x} \in \mathbb{R}^n : \xi(\underline{x}) \geq \alpha\} \\
&= \{\underline{x} \in \mathbb{R}^n : \min_{i=1(1)n} \xi_i(x_i) \geq \alpha\} \\
&= \{\underline{x} \in \mathbb{R}^n : \xi_i(x_i) \geq \alpha \quad \forall\, i = 1(1)n\} \\
&= \mathrm{X}_{i=1}^n \{x_i \in \mathbb{R} : \xi_i(x_i) \geq \alpha\} = \mathrm{X}_{i=1}^n B_\alpha(x_i^\star).
\end{aligned}$$

Lemma 2: Let $x_1^\star, \cdots, x_n^\star$ be non-precise numbers with characterizing functions $\xi_1(\cdot), \cdots, \xi_n(\cdot)$ then the function

$$\xi(x_1, \cdots, x_n) := \min_{i=1(1)n} \xi_i(x_i)$$

is a characterizing function of a non-precise vector in the sense of definition 3.

Proof: For the function $\xi(\cdot)$ $\exists\ \underline{x}_0 \in M^n$ with $\xi(\underline{x}_0) = 1$ Moreover we have to prove that the α-cuts $B_\alpha(\xi(\cdot))$ are simply connected and compact subsets of $\mathbb{R}^n$ $\forall\ \alpha \in (0,1]$. By lemma 1 we have

$$B_\alpha(\xi(\cdot)) = \mathrm{X}_{i=1}^n B_\alpha(\xi_i(\cdot))$$

and therefore $B_\alpha(\xi(\cdot))$ is closed and bounded and also simply connected.

The combination of non-precise observations $x_1^\star, \cdots, x_n^\star$ yields a non-precise element $\underline{x}^\star$ of the sample space which is described by its characterizing function.

6 Extension principle for non-precise samples

The basis for statistical inference for non-precise observations is the *non-precise combined sample element* $\underline{x}^\star$ with characterizing function $\xi(\cdot, \cdots, \cdot)$. For the generalization of functions to the situation of non-precise arguments definition 4 is used. This definition is called *extension principle* in fuzzy set theory.

In order to adapt statistical functions $\psi(x_1, \cdots, x_n)$ of data $x_1, \cdots, x_n$ to the situation of non-precise data $x_1^\star, \cdots, x_n^\star$ the non-precise combined sample element $\underline{x}^\star$ is used. The non-precise value $\psi(\underline{x}^\star)$ of the classical function $\psi(\cdot)$ at the non-precise argument value $\underline{x}^\star$ is given by its characterizing function $\eta(\cdot)$ using the characterizing function $\xi(\cdot, \cdots, \cdot)$ of $\underline{x}^\star$.

The extension principle defines $\eta(\cdot)$ by its values

$$\eta(y) := \sup_{x_i \in M} \left\{ \xi(x_1, \cdots, x_n) : \quad \psi(x_1, \cdots, x_n) = y \right\}.$$

Here the characterizing function $\xi(\cdot, \cdots, \cdot)$ of $\underline{x}^*$ is obtained from the non-precise sample by a suitable combination rule

$$\xi(x_1, \cdots, x_n) = K_n\big(\xi_1(x_1), \cdots, \xi_n(x_n)\big) \qquad \forall \quad x_i \in M.$$

In classical statistics the above considered functions $\psi(\cdot, \cdots, \cdot)$ can be

point estimators,
confidence functions,
test statistics,
general decision functions.

In Bayesian statistics the non-precise combined sample element is used to construct generalizations of

a-posteriori distributions,
Bayesian confidence regions,
non-precise predictive distributions,
Bayesian decisions based on non-precise information.

7   Classical statistical inference

7.1   Point estimators for parameters

The generalization of point estimators for statistical parameters θ of stochastic models $X \sim F_\theta$, $\theta \in \Theta$ with parameter space Θ and observation space M_X for X based on non-precise observations is possible in the following way:

An estimator $\vartheta(\cdot, \cdots, \cdot)$ for the parameter θ is a measurable function from the sample space M_X^n to Θ, i.e.

$$\vartheta : M_X^n \to \Theta.$$

For functions $\tau(\theta)$ of the parameter θ with

$$\tau : \Theta \to \Xi = \{\tau(\theta) : \theta \in \Theta\}$$

generalizations of estimators to the situation of non-precise data can also be given. In this case estimators $t(\cdot, \cdots, \cdot)$ are functions

$$t : M_X^n \to \Xi.$$

Let $t(X_1, \cdots, X_n)$ be an estimator for a transformed parameter $\tau(\theta) \in \mathbb{R}$ based on a sample $X_1, \cdots, X_n$ of a stochastic quantity X. For an observed sample $x_1, \cdots, x_n \in M_X^n$ an *estimated value*

$$\widehat{\tau(\theta)} = t(x_1, \cdots, x_n) \in \Xi$$

is obtained.

For *non-precise observations* $x_1^\star, \cdots, x_n^\star$ a reasonable generalization of estimators must yield a *non-precise estimated value* $\widehat{\tau(\theta)}^\star$ for $\tau(\theta)$.

The construction of the non-precise estimation uses the characterizing functions $\xi_i(\cdot)$ of the observations $x_i^\star$, $i = 1(1)n$ combined by a suitable combination-rule to the non-precise combined sample element $\underline{x}^\star$ of the sample space M_X^n. This non-precise combined sample element is a non-precise vector with characterizing function $\xi(\cdot, \cdots, \cdot)$ given by its values

$$\xi(x_1, \cdots, x_n) = K_n\big(\xi_1(x_1), \cdots, \xi_n(x_n)\big) \quad \text{with } x_i \in M_X.$$

as explained in section 5.

The non-precise combined sample element is the basis for the construction of a non-precise generalization of estimators for θ or functions $\tau(\theta)$ of the parameter.

Definition 6: Let $\vartheta(X_1, \cdots, X_n)$ be an estimator for the parameter θ of a stochastic model $X \sim f(. \mid \theta)$, $\theta \in \Theta$ based on a sample $X_1, \cdots, X_n$ of X. Then for non-precise observations $x_1^\star, \cdots, x_n^\star$ a non-precise estimation $\hat{\theta}^\star$ for θ based on the non-precise combined sample element $\underline{x}^\star$, whose characterizing function is $\xi(\cdot, \cdots, \cdot)$, is given by a non-precise element $\hat{\theta}^\star$ of the parameter space with characterizing function $\psi(\cdot)$, given by its values

$$\psi(\theta) := \sup \{\xi(x_1, \cdots, x_n) : \vartheta(x_1, \cdots, x_n) = \theta\}.$$

To find the supremum all elements $(x_1, \cdots, x_n)$ of the sample space have to be considered for which the condition is fulfilled.

Using the notation $\underline{x} = (x_1, \cdots, x_n)$ in short we can write

$$\psi(\theta) = \sup_{\underline{x} \in M_X^n} \{\xi(\underline{x}) : \vartheta(\underline{x}) = \theta\}.$$

Remark: The construction of the characterizing function of the non-precise estimation is an application of the generalization of classical functions given in section 4.2.

In the sample there can be also precise observations x_i. In this case the corresponding characterizing function is $I_{\{x_i\}}(\cdot)$.

7.2   Confidence regions for parameters

Let $\kappa(X_1, \cdots, X_n)$ be a confidence function with confidence level $1 - \alpha$ for a parameter θ, i.e. a function

$$\kappa : M_X^n \to \mathcal{P}(\Theta),$$

where $\mathcal{P}(\Theta)$ denotes the power set of the parameter space Θ, based on a sample $X_1, \cdots, X_n$ of the stochastic quantity $X \sim F_\theta$, $\theta \in \Theta$. Then $\kappa(X_1, \cdots, X_n)$ must fulfill

$$Probability\ \{\theta \in \kappa(X_1, \cdots, X_n)\} = 1 - \alpha \qquad \forall\ \theta \in \Theta.$$

For observed data $x_1, \cdots, x_n$ a subset $\kappa(x_1, \cdots, x_n)$ of Θ is obtained.

In case of non-precise data $x_1^\star, \cdots, x_n^\star$ a generalization of a confidence set is obtained as a *fuzzy subset* of the parameter space in the following way.

Definition 7: Let $\xi(\cdot, \cdots, \cdot)$ be the characterizing function of the non-precise combined sample element and $\kappa(X_1, \cdots, X_n)$ a confidence function. The characterizing function $\varphi(\cdot)$ of the *generalized non-precise confidence set* is given by its values

$$\varphi(\theta) := \sup \{\xi(x_1, \cdots, x_n) : \theta \in \kappa(x_1, \cdots, x_n)\}$$

where $(x_1, \cdots, x_n)$ is varying over the sample space M_X^n of X. The function $\varphi(\cdot)$ is also called membership function of the fuzzy confidence set.

Remark: For this generalized confidence set obtained by the classical confidence set $\kappa(x_1, \cdots, x_n)$ for precise data $x_1, \cdots, x_n$ the following holds

$$\varphi(\theta) = 1 \quad \text{for all} \quad \theta \in \bigcup_{(x_1, \cdots, x_n):\ \xi(x_1, \cdots, x_n)=1} \kappa(x_1, \cdots, x_n),$$

i.e. the indicator function of the union at the right hand is always below the characterizing function $\varphi(\cdot)$ of the fuzzy confidence set.

7.3   Statistical tests and non-precise data

In classical significance testing based on precise observations $x_1, \cdots, x_n$ of a stochastic quantity $X \sim F_\theta$, $\theta \in \Theta$, and observation space M_X the decision is depending on the value of a *test statistic* $T = t(X_1, \cdots, X_n)$ for sample $X_1, \cdots, X_n$ of X.

For non-precise observations $x_1^\star, \cdots, x_n^\star$ with non-precise combined sample element $\underline{x}^\star$ and corresponding characterizing function $\xi(\cdot, \cdots, \cdot)$ the value of the test statistic becomes non-precise, modelled by definition 4. The characterizing function $\psi(.)$ of this non-precise value $t^\star$ of the test statistic $T = t(x_1, \cdots, x_n)$ is given by its values

$$\psi(t) = \sup_{\underline{x} \in M_X^n} \{\xi(x_1, \cdots, x_n) : t(x_1, \cdots, x_n) = t\}.$$

For non-precise data $x_1^\star, \cdots, x_n^\star$ the value of the test statistic becomes a non-precise number $t^\star$.

If the support of $t^\star$ is a subset of the acceptance region A or a subset of the complement A^c of A then a decision on acceptance or rejection of the hypothesis is possible as for exact observations.

In case that the support of $t^\star$ has non-empty intersection with both A and A^c an immediate decision is not possible. In this case similar to sequential tests more observations are necessary.

8 Bayesian inference

8.1 Bayes' theorem for non-precise data

For continuous stochastic model $X \sim f(\cdot \mid \theta)$, $\theta \in \Theta$ and continuous parameter space Θ and a-priori density $\pi(\cdot)$ of the parameter and observation space M_X of X, Bayes' theorem for precise data $x_1, \cdots, x_n$ is

$$\pi(\theta \mid x_1, \cdots, x_n) = \frac{\pi(\theta) l(\theta; x_1, \cdots, x_n)}{\int_\Theta \pi(\theta) l(\theta; x_1, \cdots, x_n) d\theta} \qquad \forall \ \theta \in \Theta,$$

where $l(; x_1, \cdots, x_n)$ is the likelihood function. In the most simple situation of complete data the likelihood function is given by

$$l(\theta; x_1, \cdots, x_n) = \prod_{i=1}^{n} f(x_i \mid \theta) \qquad \forall \ \theta \in \Theta,$$

Remark: Using the abbreviation $\underline{x} = (x_1, \cdots, x_n)$ Bayes' theorem can be stated in the form

$$\pi(\theta \mid \underline{x}) \ \propto \ \pi(\theta) \cdot l(\theta; \underline{x}) \qquad \forall \ \theta \in \Theta,$$

where $\propto$ stands for "proportional" since the right hand of the formula is a non-normalized function which is – after normalization – a density on the parameter space Θ.

For non-precise data $D^\star = (x_1^\star, \cdots, x_n^\star)$ with corresponding characterizing functions $\xi_1(\cdot), \cdots, \xi_n(\cdot)$ the non-precise combined sample element $\underline{x}^\star$ with characterizing function $\xi(\cdot, \cdots, \cdot)$,

$$\xi : M_X^n \to [0, 1],$$

is the basis for a generalization of Bayes' theorem to the situation of non-precise data in the following way.

For all $\underline{x} \in supp(\xi(\cdot, \cdots, \cdot))$ the value $\pi(\theta \mid \underline{x})$ of the a-posteriori density $\pi(\cdot \mid \underline{x})$ is calculated using Bayes' theorem for precise data.

To every θ by variation of $\underline{x}$ in the support of the non-precise combined sample element $\underline{x}^\star$ a family

$$\left(\pi(\theta \mid \underline{x}); \ \underline{x} \in supp(\underline{x}^\star) \right)$$

of values is obtained and the characterizing function $\psi_\theta(\cdot)$ of this non-precise value is obtained via the characterizing function $\xi(\cdot, \cdots, \cdot)$ of the non-precise combined sample element by its values

$$\psi_\theta(y) := \sup \{\xi(\underline{x}) : \pi(\theta \mid \underline{x}) = y\}$$

where the supremum has to be taken over the sample space M_X^n.

Definition 8: The family $(\psi_\theta(\cdot), \theta \in \Theta)$ of non-precise values of the a-posteriori distribution is describing the imprecision of the observations $x_i^\star$, $i = 1, \cdots, n$ and is called *non-precise a-posteriori density* $\pi^\star(\cdot \mid D^\star)$, i.e.

$$\pi^\star(\cdot \mid D^\star) := (\psi_\theta(\cdot); \theta \in \Theta).$$

A graphical presentation of the non-precise a-posteriori distribution is the drawing of so-called α-level curves. These α-*level curves* are the curves which connect the ends of the α-cuts of $\psi_\theta(\cdot)$ as functions of θ.

Remark: The non-precise a-posteriori density can be used for estimations and decisions. This will be explained in the following sections.

8.2   Generalized Bayesian confidence regions

In this section a stochastic model $X \sim f(\cdot \mid \theta)$, $\theta \in \Theta$ is used and for the parameter θ an a-priori distribution $\pi(\cdot)$ is supposed to be given. The stochastic quantity describing the uncertainty about the parameter θ is denoted by $\tilde{\theta}$.

Generalizing the concept of confidence regions for non-precise data and non-precise a-posteriori distributions non-precise Bayesian confidence regions can be constructed.

For precise data $D = \underline{x} = (x_1, \cdots, x_n)$ and exact a-posteriori density $\pi(\cdot \mid \underline{x})$ a *Bayesian confidence region* Θ_0 for θ with confidence level $1 - \delta$ is defined by

$$Pr_{\pi(\cdot \mid \underline{x})}\{\tilde{\theta} \in \Theta_0\} = \int_{\Theta_0} \pi(\theta \mid \underline{x}) \, d\theta = 1 - \delta. \tag{1}$$

In case of non-precise data $D^\star = (x_1^\star, \cdots, x_n^\star)$ generalized confidence regions for θ can be constructed using the non-precise combined sample element $\underline{x}^\star$.

Definition 9: Let $\xi(\cdot, \cdots, \cdot)$ be the characterizing function of the non-precise combined sample element $\underline{x}^\star$. Then for every $\underline{x} \in supp(\xi(\cdot))$ and $1 - \delta$ a Bayesian confidence region $\Theta_{\underline{x}}$ is calculated using equation (1). The *generalized non-precise Bayesian confidence region* for θ with confidence level $1 - \delta$ is the *fuzzy subset* $\Theta^\star$ of Θ whose characterizing function $\psi(\cdot)$ is given by its values

$$\psi(\theta) := \sup_{\underline{x} \in M_X^n} \{\xi(\underline{x}) : \theta \in \Theta_{\underline{x}}\} \qquad \forall \ \theta \in \Theta.$$

Remark: Generalized Bayesian confidence regions are reasonable generalizations by the following inequality

$$I \bigcup_{\underline{x}:\ \xi(\underline{x})=1} \Theta_{\underline{x}} (\theta) \leq \psi(\theta) \qquad \forall \quad \theta \in \Theta.$$

This means that $\psi(\cdot)$ dominates the indicator functions of all classical Bayesian confidence regions $\Theta_{\underline{x}}$ with $\xi(\underline{x}) = 1$.

Highest a-posteriori density regions, abbreviated by *HPD-regions* for the parameter θ of a stochastic model $X \sim f(\cdot \mid \theta)$, $\theta \in \Theta$ with continuous parameter θ and precise data $D = \underline{x} = (x_1, \cdots, x_n)$ are defined using the a-posteriori density $\pi(\cdot \mid \underline{x})$. Generalizations to the situation of non-precise observations are possible. For details compare [14] or the monograph [15].

8.3 Non-precise predictive distributions

Information on future values of stochastic quantities X with observation space M_X and parametric stochastic model $f(\cdot \mid \theta)$, $\theta \in \Theta$ is provided by the predictive density.

In case of precise data $\underline{x} = (x_1, \cdots, x_n)$ and corresponding a-posteriori density $\pi(\cdot \mid \underline{x})$ for the parameter $\tilde{\theta}$ the predictive density $g(\cdot \mid \underline{x})$ for X conditional on data $\underline{x}$ is the conditional density of X, i.e.

$$g(x \mid \underline{x}) = \int_{\Theta} f(x \mid \theta) \pi(\theta \mid \underline{x})\, d\theta \quad \text{for all} \quad x \in M_X.$$

For non-precise data $D^* = (x_1^* \cdots, x_n^*)$ the non-precise combined sample element $\underline{x}^*$ with characterizing function $\xi(\cdot, \cdots, \cdot)$ is used for the generalization of the concept of predictive densities. This generalization is defined by a family of non-precise values for the predictive density.

Definition 10: For fixed $x \in M_X$ the vector $\underline{x}$ is varying in $supp\big(\xi(\cdot, \cdots, \cdot)\big)$ and the characterizing function $\psi_x(\cdot)$ of the *non-precise value of the predictive density* is given by its values

$$\psi_x(y) = \sup \{\xi(\underline{x}) : \ \underline{x} \in M_X^n,\ g(x \mid \underline{x}) = y\} \qquad \forall \quad x \in M_X,$$

where $g(\cdot \mid \underline{x})$ is the value of the classical predictive density based on precise data $\underline{x}$ and the supremum is to be taken over the set $supp\big(\xi(\cdot, \cdots, \cdot)\big)$. The family $\big(\psi_x(\cdot),\ x \in M_X\big)$ of non-precise values of the predictive density is called *non-precise predictive density*

$$g^*(\cdot \mid D^*) = \big(\psi_x(\cdot);\ x \in M_X\big).$$

Remark: A graphical representation of non-precise predictive densities can be given using α-level curves which are described in section 8.1.

For precise data $\underline{x} = (x_1, \cdots, x_n)$ with $\xi_i(\cdot) = I_{\{x_i\}}(\cdot)$ the resulting characterizing functions are $\psi_x(\cdot) = I_{\{g(x|\underline{x})\}}(\cdot)$. Therefore the concept is a reasonable generalization of the classical predictive density.

8.4 Non-precise a-priori distributions

Using precise a-priori distributions for parameters θ in stochastic models $X \sim F_\theta$, $\theta \in \Theta$ is a topic of critical discussions. Allowing a more general formulation of a-priori knowledge general agreement could arise on reasonable use of a-priori information on parameters.

Looking at the result in section 8.1 in natural way non-precise a-priori distributions in form of non-precise densities $\pi^\star(\cdot)$ can be used. These non-precise densities are given by the family $\pi^\star(\theta)$, $\theta \in \Theta$ of fuzzy values of the density with characterizing functions $\varphi_\theta(\cdot)$ i.e.,

$$\pi^\star(\cdot) = \big(\pi^\star(\theta); \ \theta \in \Theta\big) \ \hat{=} \ \big(\varphi_\theta(\cdot); \ \theta \in \Theta\big).$$

This formulation is also necessary to describe the sequential information gaining process which is obtained by gathering additional data. Therefore the modelling from section 8.1 has to be generalized.

Let $x_1^\star, \cdots, x_n^\star$ be n non-precise observations with corresponding characterizing functions $\xi_1(\cdot), \cdots, \xi_n(\cdot)$ and $\pi^\star(\cdot)$ a non-precise a-priori distribution with non-precise values $\pi^\star(\theta)$ and corresponding characterizing functions $\varphi_\theta(\cdot)$. Then the imprecision of the a-priori distribution has to be combined with the imprecision of the non-precise combined sample element $\underline{x}^\star$ with characterizing function $\xi(\cdot, \cdots, \cdot)$.

This combination and the generalization must yield a non-precise a-posteriori distribution for $\tilde{\theta}$.

In case of non-precise a-priori distributions formed by non-precise hyperparameters of the a-priori distribution the analysis is relatively simple. This is explained in [4] and the forthcoming monograph [15].

References

[1]   H. Bandemer (Ed.): *Modelling Uncertain Data*, Akademie Verlag, Berlin, 1993.

[2]   D. Dubois, H. Prade: Fuzzy sets and statistical data, *European Journal of Operational Research* 25, 345-356 (1986).

[3]   S. Frühwirth-Schnatter: On statistical inference for fuzzy data with applications to descriptive statistics, *Fuzzy Sets and Systems* 50, 143-165 (1992).

[4]   S. Frühwirth-Schnatter: On fuzzy Bayesian inference, *Fuzzy Sets and Systems* 60 (1993).

[5]   M.A. Gil, N. Corral, P. Gil: The minimum inaccuracy estimates in χ^2 tests for goodness of fit with fuzzy observations, *Journal of Statistical Planning and Inference* 19, 95-115 (1988).

[6]   T. Keresztfalvi: Operations on Fuzzy Numbers Extended by Yager's Family of t-Norms, in: H. Bandemer (Ed.): *Modelling Uncertain Data*, Akademie Verlag, Berlin, 1993.

[7]   R. Kruse, K.D. Meyer: *Statistics with vague data*, D. Reidel Publ., Dordrecht, 1987.

[8]   S.P. Niculescu, R. Viertl: A comparison between two fuzzy estimators for the mean, *Fuzzy Sets and Systems* 48, 341-350 (1992).

[9]   S.P. Niculescu, R. Viertl: A Fuzzy Extension of Bernoulli's Law of Large Numbers, *Fuzzy Sets and Systems* 50, 167-173 (1992).

[10] R. Viertl: Is it necessary to develop a Fuzzy Bayesian Inference, in: R. Viertl (Ed.): *Probability and Bayesian Statistics*, Plenum Press, New York, 1987.

[11] R. Viertl: Modelling of Fuzzy Measurements in Reliability Estimation, in: V. Colombari (Ed.): *Reliability Data Collection and Use in Risk and Availability Assessment*, Springer-Verlag, Berlin, 1989.

[12] R. Viertl: Statistical Inference for Fuzzy Data in Environmetrics, *Environmetrics* 1, 37-42 (1990).

[13] R. Viertl: On Statistical Inference Based on Non-precise Data, in: H. Bandemer (Ed.): *Modelling Uncertain Data*, Akademie-Verlag, Berlin, 1993.

[14] R. Viertl, H. Hule: On Bayes' theorem for fuzzy data, *Statistical Papers* 32, 115-122 (1991).

[15] R. Viertl: *Statistical Methods for Non-precise Data*, Monograph, to appear.

OPERATIVE CONTROLLING BASED ON BAYESIAN NETWORKS USING THE KALMAN FILTER TECHNIQUE

H.J. Lenz

FU Berlin, Berlin, Germany

ABSTRACT

Controlling of facts is of great importance for all kind of activities. It's objective is to check data in order to detect any kind of irregularities caused by management, employees or by environment. Given a data set, a fully specified errors-in-the-variables model and an error probability α, a statistical decision can be made whether the data are generated by the model or not.

We present the methodology of such an approach. It is mainly based on linear, Gaussian models with errors in the variables and uses generalized least-squares estimation techniques. Reparametrisation of the estimators leads to a Kalman filtering approach, cf. Schmid (1979). Inference based on statistical tests is due to Lenz, Rödel (1992). The illustrative examples are taken from Kluth (1995).

1. INTRODUCTION

A typical example of operative controlling is the following one:

A bus company is operating busses for public transport. The consumption rate x_{it} of bus i in a given period t can be measured but measurement errors must be considered. Linear aggregation (summarization) over the single consumption rates gives the total consumption rate $x_t = \Sigma\ x_{it}$ in period t. Evidently, $x_t \approx z_t$ where z_t the (absolute)

difference of the stock of fuel, i.e $I_t - I_{t-1}$, assuming no reordering in period (t-1,t]. Again z_t can be considered to be superimposed by an (independent) measurement error.
Given the data set (x_{it}, z_t) of a single period t, the linear aggregation and a fully specified error model the question arises whether irregularities are effective or not.

The corresponding model graph (Bayesian network) in this case (p=2, q=1) is simply:

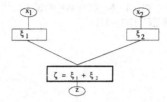

Generally speaking, model-based operative controlling is characterized by a data set $(\mathbf{x},\mathbf{z})$ and a fully specified state space model $(\mathbf{H},\mathbf{Q})$ where $\mathbf{H}$ is the observation matrix and $\mathbf{Q}$ is the variance-covariance matrix. For further details see chapter 2. Three assumptions are essential, cf. Pearl(1988), Lenz, Rödel (1992):
1. All variables are of continuous type. All interactions between variables are linear.
2. The sources of uncertainty are normally distributed. Measurement errors and errors in the equations are uncorrelated.
3. The underlying Bayesian network is not singly connected, i.e. more than one path connects any two variables.
While Pearl (1988) is voting for local computation we prefer a simultaneus approach based on a state-space model which guarantees global coherence.
The estimation technique in a slightly different context was originated by Schmid (1979). The use of multivariate test theory is due to Lenz, Rödel (1992).

2. MODELLING

The model is given by $\xi = (\xi_1, \xi_2,...,\xi_p)$ as a p-vector of true (unobservable) variables. $\mathbf{H}$ is a (q×p) matrix with q≤p and Rank($\mathbf{H}$) = q. We assume that ξ and $\zeta = \mathbf{H}\xi$ have measurement errors superimposed, i.e.

$$\mathbf{x} = \xi + \mathbf{v} \tag{1}$$
$$\mathbf{z} = \mathbf{H}\xi + \mathbf{w}. \tag{2}$$

The vectors $\mathbf{x}$, $\mathbf{z}$ are recorded.
Assume

$$E(\mathbf{v}) = E(\mathbf{w}) = 0 \text{ and } E(\mathbf{vv'}) = \mathbf{P}, E(\mathbf{ww'}) = \mathbf{R} \text{ and } E(\mathbf{vw'}) = \mathbf{Q} = \begin{pmatrix} \mathbf{P} & 0 \\ 0 & \mathbf{R} \end{pmatrix} \tag{3}$$

with Rank($\mathbf{P}$) = p and Rank($\mathbf{R}$) = q.

Stacking the vectors and matrices defined above we get

$$y = \begin{pmatrix} x \\ z \end{pmatrix}, \quad J = \begin{pmatrix} I \\ H \end{pmatrix}, \quad u = \begin{pmatrix} v \\ w \end{pmatrix} \tag{4}$$

and, consequently, the model can be written compactly as

$$y = J\xi + u . \tag{5}$$

3. ESTIMATION

Given the data y and the model (H,Q) derived from knowledge gained in the past the estimation of ξ and $\zeta = H\xi$ is straightforward. Let $\hat{\xi}$ be the generalized least squares estimator of ξ. It follows from applying the principle of generalized least squares (GLS)

$$\min_{\xi \in R^p} (y - J\xi) Q^{-1} (y - J\xi) \tag{6}$$

that

$$\hat{\xi} = (J'Q^{-1}J)^{-1} Q^{-1} J' y . \tag{7}$$

Applying Ho's matrix inversion lemma the following form of the GLS estimator can be easily derived:

$$\hat{\xi} = x + K(z - Hx) \tag{8}$$

where

$$K = P H'(HPH' + R) \tag{9}$$

is the so called gain matrix of the Kalman filter and

$$\hat{\zeta} = H\hat{\xi} . \tag{10}$$

Note, that the equations (8)-(9) represent the update formulas of the Kalman filter for state space estimation. The variance-covariance matrices of $\hat{\xi}$ and $\hat{\zeta}$ are given by

$$\Sigma_{\hat{\xi}} = P - KHP \quad \text{and} \quad \Sigma_{\hat{\zeta}} = H\Sigma_{\hat{\xi}}H'$$

Any reasoning within controlling is based on target values, realizations and deviations. Therefore, using the predicted or estimated values from (8) and (10) as target values, an inference scheme is wanted to infer about the significance of deviations between the observed values x and z and their estimates $\hat{\xi}$ and $\hat{z} = H\hat{\xi}$.

4. TESTING AND INFERENCE

The controlling itself can be easily achieved by embedding the decision about accepting or rejecting of the data (x,z) into a formal test. Using such a procedure 'large' deviations between the vectors $\hat{\xi}$ and $\mathbf{x}$ as well as between $\hat{\mathbf{z}}$ and $\mathbf{z}$ signal significant differences. If the model is

believed to be true such deviations are not coherent with the model. In other words, the model

couldn't have generated the data $(\mathbf{x},\mathbf{z})$. In the context of controlling such deviations are called

'irregularities'. They may be caused by mismanagement, carelessness, dawdling, errors in

book-keeping, suppression or even theft.

The inference or controlling process has at least two phases. The first step is to detect
'irregular' differences between target values and the corresponding realizations. The second
step concerns causal reasoning to explain the differences and effects detected from the
underlying network of conditions and causalities. We shall consider here only the monitoring
phase.

An appropriate measure of distance is the Mahalanobis distance d. Let

$$d(\hat{\xi},\mathbf{x}) = (\hat{\xi}-\mathbf{x})' \Sigma^+_{\hat{\xi}-\mathbf{x}} (\hat{\xi}-\mathbf{x}) \tag{11}$$

where $\Sigma^+_{\hat{\xi}-\mathbf{x}}$ is the Moore inverse of $\Sigma_{\hat{\xi}\cdot\mathbf{x}} = \mathbf{KHP}$.

If $\begin{pmatrix} v \\ w \end{pmatrix} \sim N(\mathbf{0},\mathbf{Q})$ then under the hypothesis that the model $(\mathbf{H},\mathbf{Q})$ is correct, $d(\hat{\xi},\mathbf{x}) \sim \chi^2_q$ and

one decides for an incoherent data set if $d(\hat{\xi},\mathbf{x}) > \chi^2_{q;1-\alpha}$ for a given probabilty of an error of

the first kind $\alpha \in (0,1)$, cf. Lenz and Rödel (1991). This inference is equivalent to saying that
the observed data set $(\mathbf{x},\mathbf{z})$ is 'too extreme' under the hypothesis that the fully specified model
$(\mathbf{H},\mathbf{Q})$ is correct.

In a similiar way the reasoning is done with respect to $\hat{\mathbf{z}}$ and $\mathbf{z}$. Under the same assumption as
above we define d as follows:

$$d(\hat{\mathbf{z}},\mathbf{z}) = (\hat{\mathbf{z}}-\mathbf{z})' \Sigma^{-1}_{\hat{\mathbf{z}}-\mathbf{z}} (\hat{\mathbf{z}}-\mathbf{z}) \sim \chi^2_q$$

where

$$\Sigma_{\hat{\mathbf{z}}-\mathbf{z}} = \mathbf{H}(\Sigma_{\hat{\xi}} + \mathbf{P})\mathbf{H'} .$$

We illustrate the approach using simulated data from a manufacturing company, cf. Kluth
(1995). The crucial role of the signal to noise ratio for operative controlling will be made
evident. The signal is linked to possible irregularities, outliers etc. whereas the noise is due to
impreciseness (measurement variance) of data and variation in the population ('natural'
variance).

5. CONTROLLING AND THE DAWDLING SCENARIO

Consider a manufacturer with a real multi-part, multi-stage and multi-task/process production. As a controlling object we take the total manhours per week and process of manufacturing. The total time x_1 of each process is computed by summing up the corresponding time slices of the various workers involved in manufacturing during a fixed week. Independently, the total can be determined from the real output per process and week. Roughly speaking, the product of the matrix T of time consumption coefficients and the output vector o per week and process gives $x_2 = T o$. Note, that T is derived from multi-moment time studies. Stacking $x=(x_1,x_2)'$, defining an isomorhic observation vector $z=0$ and, accordingly, observation matrix $H=(1,...,1,-1,...,-1)$ specifies the deterministic part of the control model used.

Dawdling can be described in a simulation model with additive or multiplicative effects. That means that dawdling increases the manhours per process either independently or dependently of the output level.

The variabilty of manhours due to 'natural' causes is modelled by a normal distribution with a mean which is dependent upon the output level. Its variance is fixed by an exogeneously given coefficient υ of variation.

The effect of dawdling in the multiplicative case is a fixed effect in the sense that the degree δ_M of dawdling takes values in the set $\{0.00, 0.01,...,0.07\}$. In the additive case random effects are considered by assuming normally distributed dawdling with time invariant mean and variance. The degree of dawdling δ_A in additive case is expressed as multiple of the mean of regular manhours consumed per process and week.

In the following we present representative results from a Monte-Carlo simulation study perfprmed by Kluth (1995). The number of simulation runs per experiment is fixed to 50 runs. The number of alarms for each of the three processes coded as A,B and C is shown in the 6 different dawdling scenarios represented in figures 1 to 6. In Fig. 1,3 and 5 the multiplicative or fixed effect is studied whereas in Fig. 2,4,6 the additive or random effect is considered. Notice that the error rates, i.e. giving blind alarm or giving unnecessarily alarm are heavily dependent upon the signal to noise ratio δ/υ, where δ is the degree of dawdling and υ is the coefficient of variation representing 'regular' variabilty of weekly manhours.

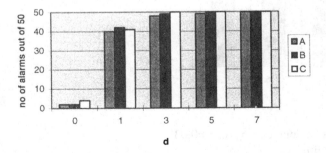

Fig. 1: Number of alarms on degree d=δ_M for υ=0.01 (dawdling type:multiplicative)

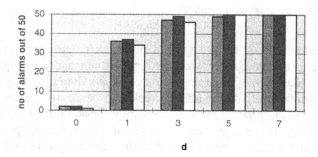

Fig. 2: Number of alarms on degree $d=\delta_A$ for $\upsilon=0.01$
(dawdling type:additive)

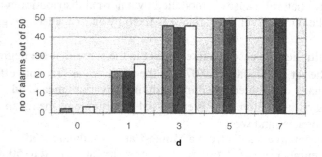

Fig. 3: Number of alarms on degree $d=\delta_M$ for $\upsilon=0.03$
(dawdling type:multiplicative)

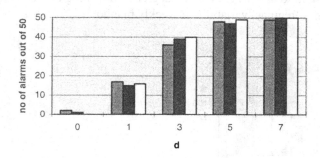

Fig. 4: Number of alarms on degree $d=\delta_A$ for $\upsilon=0.03$
(dawdling type:additive)

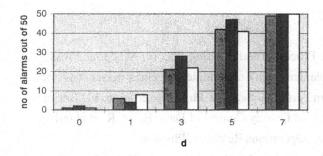

Fig. 5: Number of alarms on degree d=δ_M for υ=0.05
(dawdling type:multiplicative)

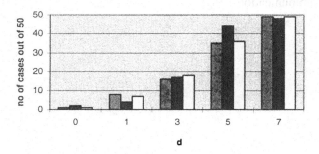

Fig. 6: Number of alarms on degree d=δ_A for υ=0.05
(dawdling type:additive)

The study makes three facts clear:
- Operative controlling must be based on a model.
- Impreciseness and randomness can be jointly modelled using an error-in-the-variables model.
- Any inference in the context of operative controlling must consider the size of the signal to noise ratio, unless the data are precise and have no natural variation. But there is a lot of empirical evidence that this would be a very weak assumption.

REFERENCES

Jazwinski, A. H. (1970) Stochastic Processes and Filtering Theory, Academic Press

Kluth, M. (1995) Konzeption, Implementation und Evaluierung eines wissensbasierten
Controllingsystems (OpCon) im Fertigungsbereich, Ph.D. dissertation, FU Berlin

Lenz, H.-J., Rödel, E. (1991) Statistical Quality Control of Data, Horst, R. et al (eds.)
Proceedings 16th Symposium on Operations Research, Physica

Pearl, Judea (1988) Probabilistic Reasoning in Intelligent Systems: Networks of Plausible
Inference, Morgan Kaufmann Publ.

Rao, C.R. (1965) Linear Statistical Inference and its Applications, Wiley

Schmid, B. (1979) Bilanzmodelle, ETH Zürich

Schneeweiß, H. (1988) Personal Communication

CONSTRUCTION, SIMULATION AND TESTING OF
CAUSAL PROBABILISTIC NETWORKS

U.G. Oppel, A. Hierle and M. Noormohammadian
University of Munich, Munich, Germany

ABSTRACT

From the probabilistic point of view a complex system subject to vagueness, randomness and uncertainty may be characterized by a multivariate probability distribution. Such a multivariate distribution may be approximated by the sequence of empirical distributions of a properly chosen sample of realizations of the system, e.g. obtained from Monte Carlo simulations of the system or by properly collected data. But how to obtain these Monte Carlo simulations?

Another way of characterizing such a system is to use an ancestral causal probabilistic network (CPN). Such a CPN is a directed graph and a family of Markov kernels. The graph describes the dependencies of the subsystems qualitatively and the Markov kernels describe them quantitatively. The conditional probabilities of the Markov kernels may be interpreted as stochastic cause–effect relations. To such a CPN a directed Markov field is associated which is the joint multivariate distribution of the system.

The representation of multivariate distribution by a CPN has some advantages: it makes even complicated multivariate distributions storable and operable, it allows for Bayesian learning by introducing and propagating of evidence, it may serve for calculation of marginal distributions, and it may be used for Monte Carlo simulations of the system. These properties make it possible to evaluate the description of the system by the multivariate distribution and the CPN. This evaluation is based on comparison of marginal distributions of the multivariate distribution with empirical distributions obtained from Monte Carlo simulations and from data. The comparison may be based on properly chosen symmetric or asymmetric distances or statistical tests. We give some examples.

1   Introduction

In many fields of application such as physics, technology, biology, medical science, and economy very often we have to deal with complex systems which are subject to randomness, uncertainty, imprecision, incompleteness and vagueness due to variabilty between and within its components or due to the kind of its observation and description. Several methods have been developed to deal with such systems.

Probability theory was the first such method and it has been successful in many fields. According to any of the many systems of axioms of probability (e.g. see Kolmogorov [11], de Finetti [5], [6], Richter [26], [27], Carnap [1], Savage [29], Stegmüller [32]) such a complex system may be described mathematically by a probability space $(\Omega, \mathfrak{K}, p)$ and the components v of such a complex system may be characterized by random variables $X_v : \Omega \to S_v$ assuming their values in rather general state spaces S_v with σ–algebras $\mathfrak{S}_v$. From the probabilistic point of view the qualitative and quantitative aspects of the total system are completely determined by the joint distribution $\mathbb{P} := p \circ X^{-1}$ of the family $X := (X_v : v \in V)$ of random variables representing its components. The joint distribution is a probability measure

$$\cdot \mathbb{P} : \mathfrak{S} \to [0,1] \quad \text{with} \quad B \mapsto \mathbb{P}(B) \; := \; p \circ X^{-1}(B) := p(X^{-1}(B))$$
$$:= \; p\big(\{\omega \in \Omega : (X_v(\omega) : v \in V) \in B\}\big)$$

on the product–σ–algebra $\mathfrak{S} := \bigotimes_{v \in V} \mathfrak{S}_v$ on the product state space $S := \prod_{v \in V} S_v$ belonging to the family of state spaces $((S_v, \mathfrak{S}_v) : v \in V)$. The system is completely determined by the probability space $(S, \mathfrak{S}, \mathbb{P})$. However, to find $\mathbb{P}$ is often very difficult. Usually $\mathbb{P}$ may be determined at most only approximately by theoretical or statistical procedures.

Very often in fields of application like medical science the systems are so extremely complicated and the information is such poor that it seems to be impossible to determine $\mathbb{P}$, neither totally nor partially. Therefore, new concepts for the description of such complex systems have been developed, for example belief, plausibility, possibility, propositional calculus, compositional systems, graphical models, neural networks, and fuzzy sets and logics; e.g. see Hajek et al. [8], Kruse et al. [12], [13] and Pearl [25]. All of these concepts and theories have some important advantages, but probability theory has some advantages, too.

The most important advantage of probability theory is that it is a rich theory. Many concepts, theorems, and procedures have been developed during its long history. To mention just a few of them: Methods of composition (integration; Fubini's and Ionescu-Tulcea's theorem) and decomposition (desintegration; Radon–Nikodym theorem, conditioning, Bayes' theorem), concepts of convergence (in probability, almost everywhere; L_p; weak, vage and uniform), inequalities (Tschebyshev's, Kolmogorov's), limit theorems (law of large numbers, ergodic theorem, Glivenko–Cantelli–Tucker theorem; central limit theorem; stationary distributions), and statistical methods (estimates for parameters or intervals of parameters; test procedures). This rich theory may be used to obtain methods for estimating the probability measure $\mathbb{P}$ from empirical knowledge, for checking $\mathbb{P}$ against data, and for simulating $\mathbb{P}$.

One way of finding $\mathbb{P}$ partially or totally is to calculate the empirical distribution from properly collected data. Here a data point x is the point of the joint state space which is obtained as the family $X(\omega) := (X_v(\omega) : v \in V)$ of states $X_v(\omega)$ of the components v for a realization ω of the complex system; i.e. $x = (X_v(\omega) : v \in V) \in S$. If we have a sequence $(x_{(k)} : k \in \mathbb{N})$ of data points which are obtained from a sequence $(X_{(k)} : k \in \mathbb{N})$ of independent (or ergodic) repetitions of X, then the law of large numbers tells us that for every $B \in S$ the relative frequency $h_k(\omega, B)$ of the occurance of B in the sequence $(X_{(k)}(\omega) : k \in \mathbb{N})$ tends with $k \to \infty$ for p–almost all $\omega \in \Omega$ to the probability $\mathbb{P}(B)$ $(= p(\{\omega \in \Omega : X(\omega) \in B\}))$; i.e.

$$h_k(\omega, B) \quad := \quad 1/k \sum_{i=1}^{k} 1_B(X_{(i)}(\omega)) \quad \to \quad \mathbb{P}(B)$$

for $k \to \infty$ for p–almost all $\omega \in \Omega$.

If we assume that all state spaces $(S_v, \mathfrak{S}_v)$ are polish spaces S_v and $\mathfrak{S}_v$ are the σ–algebras of Borel subsets and that V is countable, then S is polish, too, and $\mathfrak{S}$ the Borel–σ–algebra of S. (A topological space Z is polish, if it is separabel and completely metrizable. The σ–algebra of Borel subsets of Z is generated by the open subsets of Z.) In applications we always may make this assumption. Then the theorem of Glivenko-Cantelli (e.g. see Parthasarathy [24]) tells us that the sequence $(\mu_{k,\omega} : k \in \mathbb{N})$ of empirical distributions obtained from the sequence $(X_{(k)}(\omega) : k \in \mathbb{N})$ converges weakly to $\mathbb{P}$ for p–almost all $\omega \in \Omega$, i.e.

$$\mu_{k,\omega} \quad := \quad 1/k \sum_{i=1}^{k} \delta_{X_{(i)}(\omega)} \quad \Longrightarrow \quad \mathbb{P} \qquad \text{for } k \to \infty \text{ for } p\text{–almost all } \omega \in \Omega.$$

(A sequence $(\mu_k : k \in \mathbb{N})$ of probability measures on the polish space S converges weakly to a probability measure μ if the sequence $(\int_S f d\mu_k : k \in \mathbb{N})$ of real numbers converges to $\int_S f d\mu$ for every bounded continuous function $f : S \to \mathbb{R}$. For $x \in S$ we denote by $\delta_x : \mathfrak{S} \to [0,1]$ with $B \mapsto \delta_x(B) := 1_B(x)$ the normed Dirac measure which is concentrated in x.)

If in addition S is some finite dimensional Euklidean space $\mathbb{R}^m$, then we obtain even uniform convergence of the associated cumulative distribution functions $F_{k,\omega}(x) := \mu_{k,\omega}(\{y \in \mathbb{R}^m : y \leq x\}$ and $F(x) := \mu(\{y \in \mathbb{R}^m : y \leq x\})$, i.e.

$$\lim_{k \to \infty} \sup_{x \in \mathbb{R}^m} \left| F_{k,\omega}(x) - F(x) \right| \quad = \quad 0 \qquad \text{for } p\text{–almost all } \omega \in \Omega.$$

The empirical distribution $\mu_{k,\omega}$ is concentrated on the sample $\{X_{(1)}(\omega), \ldots, X_{(k)}(\omega)\}$ of k points $X_{(i)}(\omega)$ in the joint state space.

For some (mathematical and interpretational) purposes this "hard concentration" has disadvantages. If $S = \mathbb{R}^m$ for example, we can switch from $\mu_{k,\omega}$ to a "soft" empirical distribution by convoluting $\mu_{k,\omega}$ with some continuously distributed probability ν_k (e.g. a normal distribution) such that the sequence $(\nu_k : k \in \mathbb{N})$ converges weakly to the Dirac measure in $0 \in \mathbb{R}^m$. Because of the continuity of the convolution with respect to the weak convergence a in such a way modified theorem of the Glivenko–Cantelli type still holds.

Another way of getting more information about $\mathbb{P}$ would be to formulate some hypotheses about $\mathbb{P}$ and to apply some appropriate statistical test procedures to draw conclusions on $\mathbb{P}$. Because of the usually high dimension of the joint state space, first, it is in general not obvious how to formulate these hypotheses, and second, the usually small sample size (relative to the dimension of the joint state space) will not allow for sufficiently powerful tests. However, for lower dimensional marginal distributions this will work.

Finally, there is a way of constructing $\mathbb{P}$ using causal probabilistic networks. We shall study this method. We shall define causal probabilistic networks, point out some important properties, and present some methods for the construction, simulation and testing of causal probabilistic networks.

This will also yield a method of generating random samples of high dimensional multivariate distributions. Hence, the method of describing $\mathbb{P}$ by a causal probabilistic network will also provide us with a method for Monte Carlo simulation of $\mathbb{P}$ or of some marginal of $\mathbb{P}$ and therefore with empirical distributions which converge to $\mathbb{P}$ or its marginal, respectively.

2 Causal probabilistic networks and their associated multivariate distribution

A **causal probabilistic network (CPN)** is a directed graph $G := (V, E)$ with $E \subset V \times V$ and $(u, v) \notin E$ for $(v, u) \in E$ and a family $\mathcal{P} := (\mathcal{P}_v : v \in V)$ of Markov kernels

$$\mathcal{P}_v : S(Pa(v)) \times \mathfrak{S}_v \rightarrow [0, 1] \quad \text{with}$$

$$\big((x_u : u \in Pa(v)), B\big) \mapsto \mathcal{P}_v\big((x_u : u \in Pa(v)); B\big)$$

where $Pa(v) := \{u \in V : (u, v) \in E\}$ is the set of the parents of v in G. For $\emptyset \neq U \subset V$

$$S(U) := \prod_{u \in U} S_u \quad \text{is the product set and}$$

$$\mathfrak{S}(U) := \bigotimes_{u \in U} \mathfrak{S}_u \quad \text{is the product-σ-algebra}$$

of the family $((S_u, \mathfrak{S}_u) : u \in U)$ of state spaces $(S_u, \mathfrak{S}_u)$.

Every node $v \in V$ represents a random variable X_v with the state space $(S_v, \mathfrak{S}_v)$, E describes the dependency of $(X_v : v \in V)$ qualitatively, and $\mathcal{P}$ describes this dependency quantitatively.

(A Markov kernel $\mathcal{P}$ from a measurable space $(Y, \mathfrak{Y})$ to a measurable space $(Z, \mathfrak{Z})$ is a mapping $P : Y \times \mathfrak{Z} \to [0, 1]$ with $(y, B) \mapsto P(y; B)$ such that for fixed $y \in Y$ the mapping $P(y; \cdot) : \mathfrak{Z} \to [0, 1]$ is a probability measure and for each fixed B the mapping $P(\cdot; B) : (Y, \mathfrak{Y}) \to [0, 1]$ is measurable.)

The Markov kernel $\mathcal{P}_v$ is considered to be a family of conditional probabilities $\mathcal{P}_v((x_u : u \in Pa(v)); B)$ which are supposed to describe "the" probability of a (measurable) subset B of the state space of the node v given the state x_u of the state space of the node u for every parent node u of v. In other words, $\mathcal{P}_v$ is supposed to be "the" conditional distribution of X_v given $(X_u : u \in Pa(v))$. In applications such a Markov kernel is used to describe stochastic cause–effect relations. Therefore CPNs may be used effectively to model stochastic knowledge in expert systems; e.g. Pearl [25] and Neapolitan [18].

A very natural question now arising is the following: Is there any probability measure $\mathbb{P} : \mathfrak{S}(V) \to [0, 1]$ on the σ–algebra $\mathfrak{S}(V)$ of the joint state space of the family $(X_v : v \in V)$ of random variables X_v which is the joint distribution of the random variables X_v and which has the given Markov kernels $\mathcal{P}_v$ as conditional distributions of X_v given the values of the "parent" variables X_u? And if so, is it uniquely determined. We call a graph $G := (V, E)$ **ancestral** if there exists an enumeration $V := \{v_n : n \in N\}$ with $N \subset \mathbb{N}$ such that no descendant is enumerated before one of his ancestors. Such an enumeration we call ancestral, too.

Some examples: If G is finite and acyclic, then G is ancestral. If G is cyclic, then G is not ancestral. $\mathbb{Z}$ with the natural ordering and linear graph structure is not ancestral. The graph $G := (V, E)$ with $V := \mathbb{N} \times \mathbb{N}$ and $E := \{((k, l), (m, n)) \in V \times V : \text{either } m = k + 1 \text{ and } n = l \text{ or } m = k \text{ and } n = l + 1\}$ is ancestral.

For any CPN with an ancestral graph $G := (V, E)$ there exists a uniquely determined probability measure $\mathbb{P} : \mathfrak{S}(V) \to [0, 1]$ which is the joint distribution of the random variables X_v, has the given Markov kernels $\mathcal{P}_v$ as conditional distributions of X_v given the values of the "parent" variables X_u, and is a directed Markov field. We call this probability measure $\mathbb{P}$ the **multivariate joint probability** or the **multivariate distribution associated to this CPN**.

For finite graphs this can be shown by iterative integration:

Find an ancestral enumeration $V = \{v_i : i = 0, \ldots, n\}$ of V and define

$$S^{(i)} := S\big(\{v_j : j = 0, \ldots, i\}\big) \qquad \text{to be the product set and}$$

$$\mathfrak{S}^{(i)} := \bigotimes_{j=0}^{i} \mathfrak{S}_{v_j} \qquad \text{to be the product } \sigma\text{-algebra for } 0 \le i \le n.$$

Defining Markov kernels for $i = 0, \ldots, n$ by

$$P_i \; : \; S^{(i-1)} \times \mathfrak{S}_{v_i} \; \to \; [0,1] \quad \text{with}$$

$$\Big((x_{v_0}, \ldots, x_{v_{i-1}}), B \Big) \; \mapsto \; P_i(x_{v_0}, \ldots, x_{v_{i-1}}; B) := \mathcal{P}_{v_i}\Big((x_u : u \in Pa(v_i)); B \Big)$$

The probability measure $\mathbb{P} : \mathfrak{S}(V) \to [0,1]$ is obtained by iterative integration:

$$\mathbb{P}(A) \; := \; \int\limits_{S_{v_0}} \int\limits_{S_{v_1}} \cdots \int\limits_{S_{v_n}} 1_A(x_{v_0}, \ldots, x_{v_n}) \; P_n(x_{v_0}, \ldots, x_{v_{n-1}}; dx_{v_n}) \; \cdots$$

$$\cdots \; P_1(x_{v_0}; dx_{v_1}) \; P_0(dx_{v_0}) \qquad\qquad (2.1)$$

for $A \in \mathfrak{S} := \mathfrak{S}^{(n)} = \mathfrak{S}(V)$. For infinite graphs the existence of the probability measure $\mathbb{P} : \mathfrak{S}(V) \to [0,1]$ may be obtained from C. Ionescu–Tulcea's theorem; e.g. see Neveu [19]. Its finite dimensional marginal measures are given by iterative integrations of the form of equation 2.1.

The probability measure $\mathbb{P} : \mathfrak{S}(V) \to [0,1]$ determined by equation 2.1 does not depend on the chosen enumeration and it is a directed Markov field (in the sense of Lauritzen et al. [14]; see Oppel [20] or [22]).

The iterative construction of equation 2.1 is continuous for several norms and topologies; see Matthes et al. [17]:

Let us consider a sequence $(\mathcal{C}_n : n \in \mathbb{N})$ of CPNs $\mathcal{C}_n$ with a finite acyclic directed graph $G := (V, E)$ and sequence $(\mathcal{P}_{(n)} : n \in \mathbb{N})$ of families $\mathcal{P}_{(n)} := (\mathcal{P}_{n,v} : v \in V)$ of Markov kernels

$$\mathcal{P}_{n,v} : S(Pa(v)) \times \mathfrak{S}_v \; \to \; [0,1] \quad \text{with}$$

$$\Big((x_u : u \in Pa(v)), B \Big) \; \mapsto \; \mathcal{P}_{n,v}\Big((x_u : u \in Pa(v)); B \Big)$$

and associated multivariate distributions

$$\mathbb{P}_n : \mathfrak{S} \; \to \; [0,1]$$

and a CPN $\mathcal{C}$ with the same graph G and the family $\mathcal{P} := (\mathcal{P}_v : v \in V)$ of Markov kernels

$$\mathcal{P}_v : S(Pa(v)) \times \mathfrak{S}_v \; \to \; [0,1] \quad \text{with}$$

$$\Big((x_u : u \in Pa(v)), B \Big) \; \mapsto \; \mathcal{P}_v\Big((x_u : u \in Pa(v)); B \Big)$$

and the associated multivariate distribution

$$\mathbb{P} : \mathfrak{S} \; \to \; [0,1].$$

If the sequence $(\mathcal{P}_{(n)} : n \in \mathbb{N})$ converges uniformly to $\mathcal{P}$, i.e.

$$\lim_{n \to \infty} \|\mathcal{P}_{n,v} - \mathcal{P}_v\| = 0 \quad \text{for all } v \in V \text{ where}$$

$$\|\mathcal{P}_{n,v} - \mathcal{P}_v\| := \sup \Big\{ |\mathcal{P}_{n,v}(\vec{x}; B) - \mathcal{P}_v(\vec{x}; B)| :$$

$$(\vec{x}; B) := \big((x_u : u \in Pa(v)); B\big) \in S(Pa(v)) \times \mathfrak{S}_v \Big\},$$

then the sequence $(\mathbb{P}_n : n \in \mathbb{N})$ of associated multivariate distributions $\mathbb{P}_n$ converges uniformly to $\mathbb{P}$, i.e.

$$\lim_{n \to \infty} \|\mathbb{P}_n - \mathbb{P}\| = 0 \quad \text{where}$$

$$\|\mathbb{P}_n - \mathbb{P}\| := \sup \Big\{ |\mathbb{P}_n(A) - \mathbb{P}(A)| : A \in \mathfrak{S} \Big\}.$$

The topologies of uniform convergence of Markov kernels and of measures are very fine topologies. If all the state spaces are polish and the σ-algebras are the σ-algebras of Borel subsets, then we may consider instead of these very fine topologies the much coarser topologies of pointwise weak convergence of Markov kernels and measures.

However, if we reduce the assumption of uniform convergence of the sequences $(\mathcal{P}_{n,v} : n \in \mathbb{N})$ to the Markov kernel $\mathcal{P}_v$ to pointwise weak convergence of the sequences $(\mathcal{P}_{n,v} : n \in \mathbb{N})$ to the Markov kernel $\mathcal{P}_v$ [i.e. the weak convergence of the sequences $(\mathcal{P}_{n,v}((x_u : u \in Pa(v)); \cdot) : n \in \mathbb{N})$ of probability measures to the probability measure $\mathcal{P}_v((x_u : u \in Pa(v)); \cdot)$ for all $(x_u : u \in Pa(v)) \in S(Pa(v))$], then the weak convergence of the sequence $(\mathbb{P}_n : n \in \mathbb{N})$ of the associated multivariate distributions to $\mathbb{P}$ needs not to be true:

Let be $G := (V, E)$ with $V := \{0, 1\}$ and $E := \{(0,1)\}$, $S_0 := \{0\} \cup \{1/n : n \in \mathbb{N}\}$ with the topology induced by the usual topology on $\mathbb{R}$, $S_1 := \{0, 1\}$ with the discrete topology, and

$$\mathcal{P}_0 : \mathfrak{S}_0 \to [0, 1] \quad \text{with} \quad A \mapsto \mathcal{P}_0(A) := 1/2 \, \mu(A) + 1/2 \, \delta_0(A),$$

$$\mathcal{P}_1 : S_0 \times \mathfrak{S}_1 \to [0, 1] \quad \text{with} \quad (x_0, B) \mapsto \mathcal{P}_1(x_0; B) := \begin{cases} \delta_0(B) & \text{for } x_0 = 0, \\ \delta_1(B) & \text{for } x_0 \neq 0, \end{cases}$$

$$\mathcal{P}_{n,0} : \mathfrak{S}_0 \to [0, 1] \quad \text{with} \quad A \mapsto \mathcal{P}_{n,0}(A) := 1/2 \, \mu(A) + 1/2 \, \delta_{1/n}(A),$$

$$\mathcal{P}_{n,1} := \mathcal{P}_1, \quad \text{where} \quad \mu(\{1/n\}) := 2^{-(n+1)} \text{ for } n \in \mathbb{N}.$$

The associated multivariate distributions are $\mathbb{P} : \mathfrak{S} \to [0, 1]$ and $\mathbb{P}_n : S \to [0, 1]$ with

$$\mathbb{P}(C) := 1/2 \, \nu(C) + 1/2 \, \delta_{(0,0)}(C) \quad \text{and}$$

$$\mathbb{P}_n(C) \; := \; 1/2 \, \nu(C) + 1/2 \, \delta_{(1/n,1)}(C) \qquad \text{where}$$

$$\nu(C) \; := \; \int_{S_0} 1_C(x_0,1) \, \mu(dx_0) \qquad \text{for } C \in \mathfrak{S}.$$

Obviously, the sequences $(\mathcal{P}_{n,0} : n \in \mathbb{N})$ and $(\mathcal{P}_{n,1} : n \in \mathbb{N})$ converge pointwise weakly to $\mathcal{P}_0$ and $\mathcal{P}_1$, respectively, but the sequence $(\mathbb{P}_n : n \in \mathbb{N})$ does not converge weakly to $\mathbb{P}$.

The continuity assertion is true only under additional conditions such as: all Markov kernels are Feller kernels, have the strong Feller convergence property (**SFCP**), and have the compact uniform tightness property (**CUTP**); again see Matthes et al. [17]. For finite state spaces (each with the discrete topology) all these conditions are fulfilled and the pointwise weak convergence coincides with the uniform convergence.

What about the variation of the graph in such a limit? The links between nodes indicate dependencies. Missing links indicate the stochastic independence or the conditional stochastic independence of the variables represented by the nodes. Conditional stochastic independence means stochastic independence given the states of the latest common ancestors. Stochastic independence is measure theoretically described by product measures. Since the formation of product measures is continuous with respect to the considered topologies (of uniform or pointwise weak convergence), independence or conditional independence will be preserved in the limit. Hence, new links will not show up in the graph. Eventually, links will disappear; examples are easy to construct.

As we have seen, by iterative integration we may obtain a multivariate probability distribution from a CPN. We have mentioned some continuity properties of this iterated integration procedure. Indeed, under very mild assumptions we may obtain from a multivariate probability distribution a CPN, or even many.

Let V be countable, $\mathbb{P} : \mathfrak{S}(V) \to [0,1]$ be a probability distribution, all state spaces S_v be polish, and S_v be their Borel-σ-algebras. Choose any enumeration $V := \{v_n : n \in N\}$ with $N \subset \mathbb{N}$ of V and let $S^{(i)}$ and $\mathfrak{S}^{(i)}$ be defined as above. Furthermore, define $\pi_i : S(V) \to S^{(i)}$ and $\pi_{i,j} : S^{(i)} \to S^{(j)}$ for $0 \le j \le i \in N$ to be the canonical projections and $P_i : S^{(i-1)} \times \mathfrak{S}^{(i)} \to [0,1]$ be the desintegration kernel of the marginal measure $\mathbb{P} \circ \pi_i^{-1} : \mathfrak{S}^{(i)} \to [0,1]$ with respect to $\pi_{i,i-1}$. Applying the factorization lemma for measurable functions, we obtain from $(P_i : i \in N)$ an acyclic graph and a family of Markov kernels, i.e. we obtain a CPN. See also Oppel [20] or [22].

One important advantage of a CPN is the possibility to store a highdimensional multivariate distribution (in principle completely) which we never could store explicitly. For example, the multivariate joint distribution of a system with 1000 variables with two values each would need about 10^{300} bytes of storage.

Another important advantage of a CPN is that we may use it for Bayesian learning. Bayesian learning is deductive learning from the general knowledge (given by the multivariate distribution $\mathbb{P}$) by conditioning with the given evidence. To calculate conditional probabilities is a very simple task, in principle. If the system is very large, however, this is a very difficult task in practice. For the calculation of $\mathbb{P}$ or marginals of $\mathbb{P}$ and for introducing and propagating of evidence into such a system there are very efficient algorithms. For finite state spaces we have available the Lauritzen–Spiegelhalter algorithm (e.g. in the shell HUGIN) which is a combination of local and global backward and forward calculations based on Fubini's and Bayes' theorem; e.g. see Lauritzen et al. [15] or Jensen et al. [10]. For more general state spaces there are a number of Monte Carlo algorithms available; e.g. see Henrion [9], Chavez et al. [2], Chin et al. [3], Fung et al. [7], Shachter et al. [31]. We use variance reduced Monte Carlo methods based on importance sampling and on equation 2.1.

Unfortunately, conditioning has severe discontinuity properties with respect to all kinds of topologies. (This demands for caution in applications of expert systems with a knowledge base represented as a CPN.) For example:

Let be $G := (V, E)$ with $V := \{0, 1\}$ and $E := \{(0, 1)\}$, $S_0 := \{0\} \cup \{1/k : k \in \mathbb{N}\}$ and $S_1 := \{0, 1\}$ with the discrete topology, and

$$\mathcal{P}_0 : \mathfrak{S}_0 \to [0, 1] \quad \text{with} \quad A \mapsto \mathcal{P}_0(A) := \delta_0(A),$$

$$\mathcal{P}_1 : S_0 \times \mathfrak{S}_1 \to [0, 1] \quad \text{with} \quad (x_0, B) \mapsto \mathcal{P}_1(x_0; B) := \delta_1(B),$$

$$\mathcal{P}_{n,0} : \mathfrak{S}_0 \to [0, 1] \quad \text{with} \quad A \mapsto \mathcal{P}_{n,0}(A) := (1 - 1/n)\, \delta_{1/n}(A) + 1/n\, \delta_0(A).$$

$$\mathcal{P}_{n,1} : S_0 \times \mathfrak{S}_1 \to [0, 1] \quad \text{where} \quad (x_0, B) \mapsto \mathcal{P}_{n,1}(x_0; B) := \left\{ \begin{array}{ll} \delta_0(B) & \text{for } x_0 = 0, \\ \delta_1(B) & \text{for } x_0 \neq 0. \end{array} \right.$$

Then the sequence $(\mathcal{P}_{n,v} : n \in \mathbb{N})$ of Markov kernels $\mathcal{P}_{n,v}$ converge uniformly to $\mathcal{P}_v$ for every $v \in V$. The sequence $(\mathbb{P}_n : n \in \mathbb{N})$ of associated multivariate distributions

$$\mathbb{P}_n = (1 - 1/n) \cdot \delta_{(1/n,1)} + 1/n \cdot \delta_{(0,0)}$$

of G and $(\mathcal{P}_{n,0}, \mathcal{P}_{n,1})$ converges weakly to the associated multivariate distribution $\mathbb{P} = \delta_{(0,1)}$ of G and $(\mathcal{P}_0, \mathcal{P}_1)$, but the sequence $(Q_n(0; \cdot) : n \in \mathbb{N})$ of conditional distributions

$$Q_n(0; \cdot) := \mathbb{P}_n(\cdot \,|\, X_0 = 0) \circ X_1^{-1} = \delta_0$$

does not converge weakly to the conditional distribution

$$Q(0; \cdot) := \mathbb{P}(\cdot \,|\, X_0 = 0) \circ X_1^{-1} = \delta_1.$$

As we mentioned above, CPNs may be used effectively to model stochastic knowledge in expert systems. The knowledge base of such expert systems containing uncertain and vague knowledge may be represented by a CPN. Such a representation is most adequate especially then, if the knowledge is given or attainable in form of stochastic cause–effect relations. Stochastic cause–effect relations are usually given as conditional probabilities: if certain conditions are given, then they will imply a certain event with a certain probability. These conditional probabilities may be combined to tables of conditional probabilities, i.e. to Markov kernels. The dependencies define a directed graph.

To find the directed graph is very often not difficult. The graph structure may be found by qualitative considerations of experts. Much more difficult is the problem of finding the Markov kernels. To determine the Markov kernels in non trivial applications, many conditional probabilities have to be estimated. E.g. for the Markov kernel belonging to one node with 10 states and with 5 parent nodes having 10 states each we have to find 10^6 conditional probabilities. Usually, the available data is insufficient to estimate that many numbers. To make use of the excellent properties of CPNs, we have to design methods for the construction of the needed Markov kernels. We have to use qualitative and quantitative expert knowledge as much as possible. We have designed and applied several methods for the construction of Markov kernels based on global and local procedures.

Some of the global procedures for the construction of the family of Markov kernels for a CPN are techniques for the transformation of systems of differential equations (e.g. describing compartmental models of metabolic processes or predator–prey systems; see Oppel et al. [21], Salzsieder et al. [28], Oppel [22], [23]) and of rule–based expert systems (e.g. see Liesenfeld et al. [16]). These procedures use the deep "deterministic" knowledge and stochastify it properly to obtain a more realistic stochastic model and to be able to make use of the new powerful computational methods available for CPNs. Some of the local procedures are methods for the estimation of single Markov kernels based on functional relations, generalized linear models, and neural networks combined with properly chosen stochastifications or based on statistical estimates of (conditional) moments of the (conditional) distributions contained in the Markov kernel; e.g. see Liesenfeld et al. [16].

Nevertheless, in complex applications we are forced to choose many of the Markov kernels somewhat deliberately. Anyway, we have the problem of evaluation of the constructed CPN. For example, we have to check that the associated multivariate distribution IP or at least some of its marginals are consistent with the given data or expert knowledge. To do this, we propose to make Monte Carlo simulations of IP or its marginals, to calculate marginal distributions exactly, to calculate or to simulate "limiting" distributions (if some Markov chain may be embedded into the CPN), and to compare these results with the given data or expert knowledge by statistical testing and by measuring appropriate distances. We shall give some examples.

3   Monte Carlo simulation of a CPN and its associated multivariate distribution

Let us now consider a causal probabilistic network (CPN) with an acyclic directed graph $G := (V, E)$ with $E \subset V \times V$ and $(u, v) \notin E$ for $(v, u) \in E$ and a family $\mathcal{P} := (\mathcal{P}_v : v \in V)$ of Markov kernels

$$\mathcal{P}_v \; : \; S(Pa(v)) \times \mathfrak{S}_v \;\; \rightarrow \;\; [0, 1] \;\; \text{with}$$

$$\Big((x_u : u \in Pa(v)), B\Big) \;\; \mapsto \;\; \mathcal{P}_v\Big((x_u : u \in Pa(v)); B\Big).$$

Again for $\emptyset \neq U \subset V$

$$S(U) := \prod_{u \in U} S_u \qquad \text{is the product set and}$$

$$\mathfrak{S}(U) := \bigotimes_{u \in U} \mathfrak{S}_u \qquad \text{is the product--}\sigma\text{--algebra}$$

of the family $((S_u, \mathfrak{S}_u) : u \in U)$ of state spaces $(S_u, \mathfrak{S}_u)$, respectively. Every node $v \in V$ represents a random variable X_v with the state space $(S_v, \mathfrak{S}_v)$; we may assume that $X_v : S(V) \to S_v$ is the canonical projection. Furthermore, for $\emptyset \neq U \subset W \subset V$ we denote by

$$X_U : S(V) \;\; \rightarrow \;\; S(U) \;\; \text{with}$$

$$x := (x_v : v \in V) \;\; \mapsto \;\; X_U(x) := (x_v : v \in U) \;\; \text{and by}$$

$$X_{WU} : S(W) \;\; \rightarrow \;\; S(U) \;\; \text{with}$$

$$x_W := (x_v : v \in W) \;\; \mapsto \;\; X_{WU}(x_W) := (x_v : v \in U)$$

the canonical projections.

We shall describe a very natural method for obtaining independent (or ergodic) realizations of the random variable $X := X_V := (X_v : v \in V)$ which assumes its values in the joint state space $S (= S(V))$ and which is distributed according to $\mathbb{P}$. In other words, we shall describe a method for the Monte Carlo simulation of $\mathbb{P}$, and hence, for the Monte Carlo simulation of the CPN associated to $\mathbb{P}$. This method is similar to the one proposed by Henrion [9] for a CPN with finite state spaces and which is called the probabilistic logic sampling. (Here, we do not want to propagate evidence in a CPN via Monte Carlo simulation. For that purpose there are more efficient algorithms; e.g. see Shachter–Peot [31].)

The proposed Monte Carlo simulation methods works as follows:
First, find some (appropriate) ancestral enumeration $V := \{v_i : i = 0, \ldots, n\}$ of V.

Again, define

$$S^{(i)} := S\big(\{v_j : j = 0, \ldots, i\}\big) \qquad \text{to be the product set and}$$

$$\mathfrak{S}^{(i)} := \bigotimes_{j=0}^{i} \mathfrak{S}_{v_j} \qquad \text{to be the product } \sigma\text{-algebra for } 0 \leq i \leq n.$$

Again, defining Markov kernels for $i = 0, \ldots, n$ by

$$P_i \; : \; S^{(i-1)} \times \mathfrak{S}_{v_i} \; \rightarrow \; [0,1] \quad \text{with}$$

$$\big((x_{v_0}, \ldots, x_{v_{i-1}}), B\big) \; \mapsto \; P_i(x_{v_0}, \ldots, x_{v_{i-1}}; B) := \mathcal{P}_{v_i}\big((x_u : u \in Pa(v_i)); B\big)$$

The joint distribution $\mathbb{P} : \mathfrak{S}(V) \rightarrow [0,1]$ associated to the given CPN is obtained by iterative integration:

$$\mathbb{P}(A) \; := \; \int_{S_{v_0}} \int_{S_{v_1}} \cdots \int_{S_{v_n}} 1_A(x_{v_0}, \ldots, x_{v_n}) \; P_n(x_{v_0}, \ldots, x_{v_{n-1}}; dx_{v_n}) \; \cdots$$

$$\cdots \; P_1(x_{v_0}; dx_{v_1}) \; P_0(dx_{v_0}) \qquad\qquad (3.2)$$

for $A \in \mathfrak{S} := \mathfrak{S}^{(n)} = \mathfrak{S}(V)$. This equation is the basis for a Monte Carlo simulation of $\mathbb{P}$. A Monte Carlo simulation of $\mathbb{P}$ (i.e. of the associated acyclic CPN) yields a point $x := (x_v : v \in V) \in S(V)$ which is a random realization of the family of random variables $(X_v : v \in V)$ on $(S(V), \mathfrak{S}(V), \mathbb{P})$.

The flow chart for the Monte Carlo simulation of $\mathbb{P}$ is based on equation 3.2:

Let $V := \{v_i : i = 0, \ldots, n\}$ be the chosen ancestral enumeration of V.

0. Choose $x_{v_0} \in S_{v_0}$ at random according to the probability distribution $P_0 = \mathcal{P}_{v_0}$. (The Markov kernel P_0 is degenerate!)

1. Choose $x_{v_1} \in S_{v_1}$ at random according to the probability distribution $P_1(x_{v_0}; \cdot)$.

2. Choose $x_{v_2} \in S_{v_2}$ at random according to the probability distribution $P_2(x_{v_0}, x_{v_1}; \cdot)$.

$$\vdots$$

n. Choose $x_{v_n} \in S_{v_n}$ at random according to the probability distribution $P_n(x_{v_0}, \ldots, x_{v_{n-1}}; \cdot)$.

Flow chart:

$$\xrightarrow{P_0} \; x_{v_0} \; \xrightarrow{P_1(x_{v_0}; \cdot)} \; x_{v_1} \; \xrightarrow{P_2(x_{v_0}, x_{v_1}; \cdot)} \; x_{v_2} \; \xrightarrow{P_3(x_{v_0}, x_{v_1}, x_{v_2}; \cdot)} \; x_{v_3} \; \longrightarrow \; \cdots$$

Remember: $P_k(x_{v_0}, \dots, x_{v_{k-1}}; \cdot) = \mathcal{P}_{v_k}\Big((x_u : u \in Pa(v_k)); \cdot\Big).$

Let us now assume that every state space S_v is polish and $\mathcal{S}_v$ is its Borel-σ-algebra and that $(x_{(m)} : m \in \mathbb{N})$ with $x_{(m)} := (x_v^{(m)} : v \in V) \in S(V)$ is a sequence of independent (or ergodic) repetitions of Monte Carlo simulations of $\mathbb{P}$. From the Glivenko–Cantelli theorem we know:

$$\mathbb{P}^{\xi,k} := 1/k \sum_{m=1}^{k} \delta_{x_{(m)}} \implies \mathbb{P} \quad \text{for } k \to \infty$$

i.e. the empirical distributions $\mathbb{P}^{(\xi,k)}$ associated to the sequence $\xi := (x_{(m)} : m \in \mathbb{N})$ of independent (or ergodic) repetitions of Monte Carlo simulations of $\mathbb{P}$ converge weakly to $\mathbb{P}$ for $\mathbb{P}^{\mathbb{N}}$–almost all such repetitions.

Analogue statements are true for marginal distributions $\mathbb{P} \circ X_W^{-1}$ with $\emptyset \neq W \subset V$ of $\mathbb{P}$.

An example of a CPN:
The CPN "NEPHRO_RISK" for the risk of progression of the diabetic nephropathy.

To have a realistic CPN in mind, let us have a look at the CPN "NEPHRO_RISK". This CPN was designed for the prognosis of the risk of progression of diabetic nephropathy. It was constructed in a close cooperation by the mathematician A. Hierle and the physicians B. Liesenfeld and H.–J. Lüddeke as a part of the DIADOQ project "Diabetes mellitus: Optimierte Betreuung durch wissensbasierte Qualitätssicherung" of the research program MEDWIS sponsored by the German Minister of Science and Technology (BMFT).

The nodes of the CPN "NEPHRO_RISK" are representing the following variables:

age, anamnesis, ACE, smoking, HbA 1c;
albumin: 2 years ago, today;
blood pressure (Riva–Rocci): 2 years ago, today;
alpha–1–microglobulin: 2 years ago, today;
risk of progression of diabetic nephropathy.

The selection of the variables and the structure of the graph of this CPN was the result of a balancing consideration taking into account the possibly important dependencies, the physician's availability of the data, and the attainability of knowledge needed for the determination of the Markov kernels.

The quantitative knowledge needed for the Markov kernels was "extracted" from medical experts and data pools of clinics. The method was based on a linear model describing expected values of nodes given the states of the parent nodes as a weighted linear combination of scored expected values given the state of the single parent node. The conditional distribution was obtained by a stochastification with a properly chosen and discretized normal distribution centered at the conditional expectation given the states of the parent nodes. For more details see Liesenfeld et al. [16]; a detailed description of the procedure will be published soon.

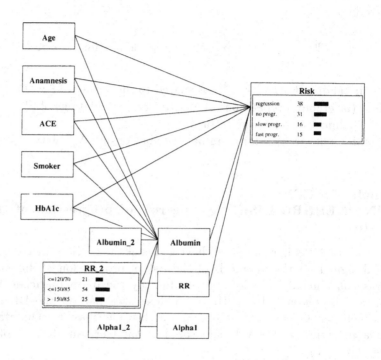

Figure 1: The directed graph of CPN "NEPHRO_RISK". The windows of the nodes "RR_2" and "Risk" are open and displaying the distributions of their variables.

As a result of the construction procedure described above, a first version of this CPN was obtained. It was checked by physicians comparing their experience with the prognosis of the HUGIN driven expert system based on this CPN. It passed this type of evaluation. Furthermore, this CPN was checked using data from the data pools of a clinic using statistical methods (such as L_1- and L_2-norms, Kullback–Leibler divergence, and chi–square fitting tests; see below). The first version of this CPN failed this statistical evaluation. Of course, the available data was at least as questionable as the experience of the physicians. But we had the feeling, that we should adapt the CPN more to the data without loosing the acception by the physicians.

Such a CPN is a very complex and delicate system. It is like a mobile: if you touch it somewhere slightly, it may change far away wildly. Hence, any changes have to be supervised by properly designed controlling. To adapt the constructed CPN "NEPHRO_RISK" to data, a random search optimization procedure was applied. The scores and, hence, the Markov kernels were changed at random (by a properly chosen selection probability). Then the L_2-norm distance between some marginals of the old associated multivariate distribution and the same marginals of the empirical distribution of the data was compared with the L_2-norm distance between the marginals of the new associated multivariate distribution and the marginals of the empirical distribution of the data. If the random variation of the CPN resulted in an improvement, this variation was taken as the starting point for the next step. If not, the old CPN was taken as the starting point again. Indeed, this procedure resulted in some improvement. But we are not completely satisfied until now. To obtain better results we designed Monte Carlo simulations for CPNs and methods for calculation, simulation and comparison of marginal distributions of the associated multivariate distributions of CPNs. Below, we shall describe such methods and report on some results.

Figure 1 shows the graph of the CPN "NEPHRO_RISK". The windows of two nodes are displaying their onedimensional marginal distributions of the associated multivariate distribution. For sake of simplicity, we shall only consider these two nodes and their one- and twodimensional distributions.

Examples of Monte Carlo simulations:
Simulations of a twodimensional marginal distribution of the associated multivariate distribution of the CPN "NEPHRO_RISK".

For the CPN "NEPHRO_RISK" we consider the joint distribution of the two variables "RR_2" and "Risk". The variable "RR_2" has the three states

$$1 \ : \ \text{"}\leq 120/70\text{"},$$
$$2 \ : \ \text{"}120/70 < \ldots \leq 150/85\text{"},$$
$$3 \ : \ \text{"}> 150/85\text{"} \ [mm \ Hg];$$

it is describing the Riva–Rocci blood pressure of the patient two years ago. The variable "Risk" is describing the risk of progression of diabetic nephropathy; it has the four states

$$1 \ : \ \text{"regression"},$$
$$2 \ : \ \text{"no progression"},$$
$$3 \ : \ \text{"slow progression"},$$
$$4 \ : \ \text{"fast progression"}.$$

Table 1 to 5 give examples of Monte Carlo simulations of the CPN "NEPHRO_RISK". They show the empirical probabilities (which here are the relative frequencies), the relative errors in percent (which here are the quotients of the standard deviation and the expectation of the sample divided by the squareroot of the size of the sample), and the absolute frequencies (counts) for different sizes of samples.

RR_2	Risk	probability	rel. error in %	counts
1	1	0.0760	22.1	19
2	1	0.2240	11.8	56
3	1	0.0480	28.2	12
1	2	0.0560	26.0	14
2	2	0.2040	12.5	51
3	2	0.0520	27.1	13
1	3	0.0200	44.4	5
2	3	0.0920	19.9	23
3	3	0.0800	21.5	20
1	4	0.0200	44.4	5
2	4	0.0760	22.1	19
3	4	0.0520	27.1	13

Table 1: **250** Monte Carlo simulations of the CPN "NEPHRO_RISK".

RR_2	Risk	probability	rel. error in %	counts
1	1	0.1060	9.2	106
2	1	0.2140	6.1	214
3	1	0.0550	13.1	55
1	2	0.0660	11.9	66
2	2	0.1720	6.9	172
3	2	0.0720	11.4	72
1	3	0.0160	24.8	16
2	3	0.0920	9.9	92
3	3	0.0630	12.2	63
1	4	0.0100	31.5	10
2	4	0.0760	11.0	76
3	4	0.0580	12.8	58

Table 2: **1000** Monte Carlo simulations of the CPN "NEPHRO_RISK".

RR_2	Risk	probability	rel. error in %	counts
1	1	0.1197	2.7	1197
2	1	0.2083	1.9	2083
3	1	0.0465	4.5	465
1	2	0.0616	3.9	616
2	2	0.1672	2.2	1672
3	2	0.0865	3.2	865
1	3	0.0183	7.3	183
2	3	0.0843	3.3	843
3	3	0.0551	4.1	551
1	4	0.0108	9.6	108
2	4	0.0743	3.5	743
3	4	0.0674	3.7	674

Table 3: **10000** Monte Carlo simulations of the CPN "NEPHRO_RISK".

RR_2	Risk	probability	rel. error in %	counts
1	1	0.1235	0.8	12346
2	1	0.2049	0.6	20492
3	1	0.0471	1.4	4709
1	2	0.0605	1.2	6054
2	2	0.1702	0.7	17017
3	2	0.0848	1.0	8483
1	3	0.0193	2.3	1926
2	3	0.0885	1.0	8848
3	3	0.0535	1.3	5347
1	4	0.0096	3.2	958
2	4	0.0704	1.1	7036
3	4	0.0678	1.2	6784

Table 4: **100000** Monte Carlo simulations of the CPN "NEPHRO_RISK".

RR_2	Risk	probability	rel. error in %	counts
1	1	0.1226	0.3	122604
2	1	0.2067	0.2	206731
3	1	0.0477	0.4	47667
1	2	0.0610	0.4	60951
2	2	0.1695	0.2	169477
3	2	0.0838	0.3	83799
1	3	0.0189	0.7	18913
2	3	0.0901	0.3	90054
3	3	0.0531	0.4	53111
1	4	0.0100	1.0	9970
2	4	0.0689	0.4	68943
3	4	0.0678	0.4	67780

Table 5: **1000000** Monte Carlo simulations of the CPN "NEPHRO_RISK".

4 Construction of marginal distributions of the associated multivariate distribution of a CPN

Let $\mathbb{P} : \mathfrak{S}(V) \to [0,1]$ be the associated multivariate distribution of a given CPN with the graph $G := (V, E)$ and the family of Markov kernels $(\mathcal{P}_v : v \in V)$. Again we assume that all state spaces are polish.

For $v \in V$ and $\emptyset \neq W \subset V$ with $v \notin W$ the conditional distribution of X_v given $X_W := (X_w : w \in W)$ exists and is denoted by $\mathcal{P}_{v|W} : S(V) \times \mathfrak{S}_v \to [0,1]$. We have:

1. For fixed $B \in \mathfrak{S}_v$ the function $\mathcal{P}_{v|W}(\,\cdot\,; B) : \big(S(V), X_W^{-1}(\mathfrak{S}(W))\big) \to [0,1]$ with $x \mapsto \mathcal{P}_{v|W}(\,\cdot\,; B)$ is measurable.

2. For each $B \in \mathfrak{S}_v$ and $C \in \mathfrak{S}(W)$ we have

$$\mathbb{P}\big(X_v^{-1}(B) \cap X_W^{-1}(C)\big) = \int\limits_{X_W^{-1}(C)} \mathcal{P}_{v|W}(x; B) \; \mathbb{P}(dx).$$

Applying the factorization lemma, we get a function $\mathbb{P}_{v|W} : S(W) \times \mathfrak{S}_v \to [0,1]$ such that $\mathcal{P}_{v|W}(x; B) = \mathbb{P}_{v|W}(X_W(x), B)$ for all $x \in S(V)$ and all $B \in \mathfrak{S}_v$.

The Markov kernel $\mathbb{P}_{v|W} : S(W) \times \mathfrak{S}_v \to [0,1]$ is a desintegration kernel of the marginal distribution

$$\mathbb{P}_U := \mathbb{P} \circ X_U^{-1} \qquad \text{for } U := W \cup \{v\}$$

with respect to the projection $X_{UW} : S(U) \to S(W)$; i.e. we have:

$$\mathbb{P}_U(C) = \int\limits_{S(W)} \int\limits_{S_v} 1_C(x_W, x_v) \; \mathbb{P}_{v|W}(x_W, dx_v) \; \mathbb{P}_W(dx_W)$$

where $x_W := (x_u : u \in W)$. For discrete state spaces $\mathbb{P}_{v|W}$ may be calculated easily via HUGIN (and API) and $\mathbb{P}_U$ may be calculated iteratively via HUGIN + API in combination with additional programs. Future versions of HUGIN will contain such algorithms.

For general state spaces $\mathbb{P}_U$ may be determined approximately via Monte Carlo simulation and $\mathbb{P}_{v|W}$ may be determined approximately via variance reduced Monte Carlo simulation.

Let us now describe a method to calculate $\mathbb{P}_{v|W}$ using HUGIN: For discrete state spaces we have $\mathbb{P}_{v|W}((x_w : w \in W); B) = \mathbb{P}(X_v \in B | X_w = x_w$ for $w \in W)$, and hence:

1. Select $(x_w : w \in W) \in S(W)$.

2. Open the windows of each of the nodes $w \in W$.

3. Introduce evidence "$X_w = x_w$" for all $w \in W$ in any order and propagate.

4. Open the window of node v: the distribution displayed at this window is $\mathbb{P}_{v|W}\big((x_w : w \in W); \cdot\big)$.

This procedure may be implemented as an automatic procedure using the HUGIN API. Let us now describe a method to calculate $\mathbb{P}_U$ iteratively using HUGIN+API and additional programs:

1. For $W := \{w\}$ the onedimensional marginal $\mathbb{P}_W$ is displayed at the window of node w.

2. For $W := \{w_1, w_2\}$ calculate the desintegration kernel $\mathbb{P}_{w_2,\{w_1\}}$ and take the marginal distribution $\mathbb{P}_{\{w_1\}}$ from the window of node w_1.
   Obtain $\mathbb{P}_{\{w_1, w_2\}}$ from:

$$\mathbb{P}_{\{w_1, w_2\}}(C) = \int\limits_{S(\{w_1\})} \int\limits_{S_{w_2}} 1_C(x_{w_1}, x_{w_2}) \; \mathbb{P}_{w_2,\{w_1\}}(x_{w_1}; dx_{w_2}) \; \mathbb{P}_{\{w_1\}}(dx_{w_1})$$

(which is a finite sum only). (4.3)

$\vdots$

$k+1$. For $W := \{w_1, w_2, \ldots, w_{k+1}\}$ calculate the desintegration kernel $\mathbb{P}_{w_{k+1},\{w_1,\ldots,w_k\}}$ and take the marginal distribution $\mathbb{P}_{\{w_1,\ldots,w_k\}}$ from the preceeding calculation.
   Obtain $\mathbb{P}_{\{w_1,\ldots,w_{k+1}\}}$ from $\mathbb{P}_{\{w_1,\ldots,w_k\}}$ and $\mathbb{P}_{w_{k+1},\{w_1,\ldots,w_k\}}$ analogue to equation 4.3.

Example: Calculation of a twodimensional joint distribution

We take the CPN "NEPHRO_RISK" designed for the prognosis of the risk of progression of diabetic nephropathy which has been introduced above. We shall show how to calculate the joint distribution of the following two variables:

RR_2   := (Riva–Rocci–) blood pressure two years ago
Risk   := risk of progression of diabetic nephropathy

This calculation will be done in three steps:

1. Calculation of onedimensional marginal distribution of "RR_2".

2. Calculation of conditional probabilities of "Risk" given "RR_2".

3. Summation of the products of marginal and conditional probabilities.

a)   Calculation of onedimensional marginal distribution of "RR_2"

We use HUGIN to calculate the probability distribution of the variable "RR_2". This distribution is the onedimensional marginal distribution of the multivariate distribution which is associated to the CPN "NEPHRO_RISK". The algorithm of HUGIN will calculate all the onedimensional marginal distributions of the associated multivariate distribution. To get to know it, we only have to open the window of "RR_2". This window will display this distribution immediately graphically. (The precise numbers of this distribution may be obtained by using the program HUGIN–API which may be accessed by the programming language ANSI C.) This is shown in Figure 1. The window of "Risk" displays the distribution of the variable "Risk". This distribution again is a onedimensional marginal of the multivariate distribution associated to the CPN "NEPHRO_RISK".

b)   Calculation of conditional probabilities of "Risk" given "RR_2"

We use HUGIN to introduce each of the possible three states of the variable "RR_2" and to propagate this evidence. To do this, we open the window of "RR_2" and the window of "Risk". Then we set one of the states of the window of "RR_2"; this state will show the 100% bar. After propagation the window of "Risk" will display the conditional distribution of "Risk" given the introduced evidence in the window of "RR_2".

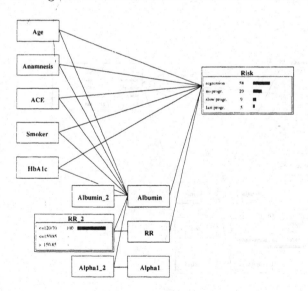

Figure 2: Introduce evidence "RR_2 ≤ 120/70" and propagate: Now the window of "Risk" displays the conditional distribution of "Risk" given "RR_2 ≤ 120/70".

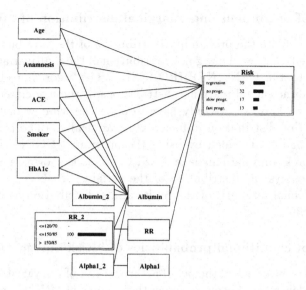

Figure 3: Introduce evidence "120/70 < RR_2 ≤ 150/85" and propagate: Now the window of "Risk" displays the conditional distribution of "Risk" given "120/70 < RR_2 ≤ 150/85".

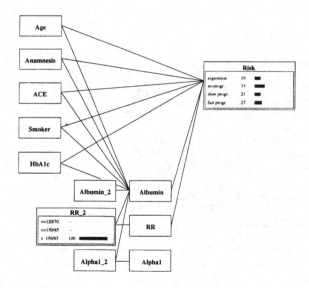

Figure 4: Introduce evidence "150/85 < RR_2" and propagate: Now the window of "Risk" displays the conditional distribution of "Risk" given "150/85 < RR_2".

c)   Summation of the products of marginal and conditional probabilities

The summation of the products of the marginal and the conditional probabilities results in the ("calculated") twodimensional joint distribution of the variables "RR_2" and "Risk". We call it the exact twodimensional joint distribution of the variables "RR_2" and "Risk" obtained from the CPN "NEPHRO_RISK".

RR_2	Risk	probability
1	1	0.12251
2	1	0.20676
3	1	0.04734
1	2	0.06098
2	2	0.16915
3	2	0.08408
1	3	0.01876
2	3	0.08999
3	3	0.05337
1	4	0.00988
2	4	0.06945
3	4	0.06772

Table 6:   The exact twodimensional joint distribution of the variables "RR_2" and "Risk" obtained from the CPN "NEPHRO_RISK".

5   Methods for evaluation of a CPN

Let us consider a CPN with an acyclic finite graph $G := (V, E)$ and a nonempty subset W of V. The associated multivariate distribution $\mathbb{P} : \mathfrak{S}(V) \to [0,1]$ of this CPN has the marginal distribution $\mathbb{P}_W := \mathbb{P} \circ X_W^{-1} : \mathfrak{S}(W) \to [0,1]$. Furthermore, let

$$\mathbb{P}^{\xi,k} := 1/k \sum_{m=1}^{k} \delta_{x^{(m)}} : \mathfrak{S}(V) \to [0,1]$$

be the empirical distribution associated to the sequence $\xi := (x^{(m)} : m \in \mathbb{N})$ obtained from independent or ergodic repetitions of Monte Carlo simulations of $\mathbb{P}$ or from data. From the Glivenko-Cantelli theorem we know that for independent or ergodic repetitions of Monte Carlo simulations of $\mathbb{P}$ the sequence of empirical distributions converges weakly to $\mathbb{P}$ with probability one. If the data is collected properly, we may assume this convergence, too. (Unfortunately, very often the data is not collected properly. In many medical data pools we have strongly dependent repetitions; e.g. one person produces many data sets which are not recognizable as being generated by the same person. In such a case we should reduce the weight of the data corresponding to the multiplicity.) If the sequence of empirical distributions converges weakly to $\mathbb{P}$, then also the sequence $(\mathbb{P}_W^{\xi,k} : k \in \mathbb{N})$ of the marginals

$$\mathbb{P}_W^{\xi,k} \; := \; \mathbb{P}^{\xi,k} \circ X_W^{-1} : \mathfrak{S}(W) \to [0,1]$$

of $\mathbb{P}^{\xi,k}$ converges weakly to the marginal $\mathbb{P}_W$ of $\mathbb{P}$. We shall now discuss some methods for the evaluation of a CPN by evaluating instead of its associated multivariate distribution some of its marginal distributions. Because of very frequent lack of data we may only evaluate those marginal distributions for which enough data is available. We propose to apply symmetric measures of distance like the norm of total variation, L_p-norms or other metrics (which coincide for finite state spaces) or asymmetric measures of similarity or coincidence like the Kullback–Leibler divergence or some test for fitting (e.g. Kolmogorov, Kolmogorov–Smirnov, chi–square fitting).

We may compare exact marginal distributions with empirical distributions obtained from data or we may compare empirical distributions obtained from Monte Carlo simulations of the associated multivariate distribution with empirical distributions obtained from data.

Now, we shall demonstrate these procedures by applying them to our example, the CPN "NEPHRO_RISK". Again, we shall consider the two variables "RR_2" and "Risk" and their twodimensional joint distribution. We compare this distribution to the ones obtained by Monte Carlo simulation and from data. We also compare these with each other. The comparison is based on the L_1-norm, the L_2-norm, the Kullback–Leibler divergence and the chi–square fitting. The first variable has three states, the second variable has four states, and hence, the joint state space has 12 states. If we take the counting measure on this joint state space, the L_p-norm of the density f_μ of some measure μ on the joint state space with respect to this counting measure will be

$$\|\mu\|_p \; := \; \|f_\mu\|_p \; := \; \left(\sum_{i=1}^{3} \sum_{j=1}^{4} \left| \mu\big(\{(i,j)\}\big) \right|^p \right)^{1/p}.$$

The (symmetric) L_p-distance of two measures μ and ν is defined by $\|\mu - \nu\|_p$. If two measures μ and ν are positive for each singleton, the (asymmetric) Kullback–Leibler divergence from μ to ν is defined by

$$D_{KL}(\mu;\nu) \; := \; \sum_{i=1}^{3} \sum_{j=1}^{4} \nu\big(\{(i,j)\}\big) \cdot \log\left(\nu\big(\{(i,j)\}\big) \big/ \mu\big(\{(i,j)\}\big) \right).$$

The (asymmetric) chi–square estimate for (the unsymmetric) testing of a measure μ against an empirical distribution ν belonging to a sample of size k is

$$D_{\chi^2}(\mu;\nu) := k \cdot \sum_{i=1}^{3} \sum_{j=1}^{4} \left(\nu\big(\{(i,j)\}\big) - \mu\big(\{(i,j)\}\big) \right)^2 / \mu\big(\{(i,j)\}\big).$$

Table 8 shows some of the results of the comparison of one distribution against the other distribution. We write for short:

μ_1 := exact joint distribution,

μ_2 := empirical distribution obtained from 250 MC simulations,

μ_3 := empirical distribution obtained from 1.000 MC simulations,

μ_4 := empirical distribution obtained from 1.000.000 MC simulations,

μ_5 := empirical distribution obtained from 264 patients with repetitions,

μ_6 := empirical distribution obtained from 53 patients without repetitions.

Finally, we denote by "μ_i/μ_j" the "... distance from μ_i to μ_j" or "μ_i against μ_j".

RR_2	Risk	from 264 pairs	from 53 pairs
1	1	0.193	0.1698
2	1	0.197	0.2075
3	1	0.03	0.0377
1	2	0.133	0.1321
2	2	0.121	0.1509
3	2	0.015	0.0189
1	3	0.034	0.0189
2	3	0.057	0.0566
3	3	0.019	0.0189
1	4	0.027	0.0189
2	4	0.098	0.0943
3	4	0.076	0.0755

Table 7: The twodimensional empirical joint distributions of the variables "RR_2" and "Risk" obtained from 264 pairs of data of not necessarily different patients and from 53 pairs of data of different patients, respectively.

We had available a set of 264 pairs of data of "RR_2" and "Risk". This set was taken from the data pool of the clinic Krankenhaus München Bogenhausen. The empirical distribution of this sample is presented in Table 7. A careful inspection of these pairs of data revealed that the 264 pairs of data came from 53 patients only. Hence, a large part of this sample is strongly dependent. At random we selected only one pair of data for each of the 53 patients. The empirical distribution of this sample is presented in Table 7.

	L_1–distance	L_2–distance	KL–divergence	chi–square	sample size k
μ_1/μ_2	0.1985	0.0763	0.0350	—	—
μ_1/μ_3	0.0821	0.0283	0.0051	—	—
μ_1/μ_4	0.0024	0.0008	-0.0001	—	—
μ_1/μ_5	0.4233	0.1460	0.1733	—	—
μ_1/μ_6	0.3218	0.1226	0.1296	—	—
μ_2/μ_1	0.1985	0.0763	0.0336	16.7444	250
μ_2/μ_3	0.1460	0.0548	0.0188	9.7702	250
μ_2/μ_4	0.1983	0.0763	0.0335	16.7016	250
μ_2/μ_5	0.522	0.1875	0.2066	128.9095	250
μ_2/μ_6	0.4234	0.1573	0.1603	101.7737	250
μ_5/μ_1	0.4233	0.1460	0.1460	77.0654	264
μ_5/μ_2	0.522	0.1875	0.1989	117.6430	264
μ_5/μ_6	0.104	0.0438	0.0127	7.3118	264
μ_6/μ_1	0.3218	0.1226	0.1078	11.0418	53
μ_6/μ_2	0.4234	0.1573	0.1518	17.6455	53
μ_6/μ_5	0.104	0.0438	0.0119	1.2198	53

Table 8:   A comparison of symmetric and asymmetric distances between the twodimensional distributions obtained from the CPN "NEPHRO_RISK" and from clinical data pools.

From Table 8 we learn that the missing independence within the sample from the 264 patients causes severe problems: The hypotheses that the CPN "NEPHRO_RISK" is true will be rejected for reasonable levels of significance (e.g. $\alpha = 0.01$, 0.05 and 0.10). This is also suggested by the distance measures (L_1, L_2, KL). The opposite is true for the independent sample from the 53 patients.

Acknowledgement: This research was supported by the German Minister of Science and Technology (BMFT) and done as part of the project "DIADOQ: Diabetes Mellitus. optimierte Betreuung durch wissensbasierte Qualitätssicherung".

References

[1] Carnap, R.: Logical Foundations of Probability. The University of Chicago Press: 1950.

[2] Chavez, R.M.; Cooper, G.F.: An empirical evaluation of a randomized algorithm for probabilistic inference. In: Henrion, M.; Shachter, R.D.; Kanal, L.N.; Lemmer, J.F. (eds.): Uncertainty in Artificial Intelligence 5. Elsevier Science Publishers B.V. (North–Holland), 1990.

[3] Chin, H.L.; Cooper, G.F.: Bayesian belief network inference using simulation. In: Kanal, L.N.; Levitt, T.s.; Lemmer, J.F. (eds.): Uncertainty in Artificial Intelligence 3. Elsevier Science Publishers B.V. (North–Holland), 1989.

[4] Feller, W.: An Introduction to Probability Theory and its Applications. J. Wiley: New York, 1950. Volume I. J. Wiley and Sons: New York, 1960. Volume II. J. Wiley and Sons: London, 1966.

[5] Finetti, B. de: Probability: Sulle funzioni a incremento aleatorio. Rend. Acad. Lincei Cl. Sci. Fis. Mat. 10 (6), 1929.

[6] Finetti, B. de: Theory of Probability. Volume 1. J. Wiley and Sons: London–New York–Sydney–Toronto, 1974.

[7] Fung, R.; Chang, K.-C.: Weighing and integrating evidence for stochastic simulation in Bayesian networks. In: Henrion, M.; Shachter, R.D.; Kanal, L.N.; Lemmer, J.F. (eds.): Uncertainty in Artificial Intelligence 5. Elsevier Science Publishers B.V. (North–Holland), 1990.

[8] Hájek, P.; Havránek, T.; Jirousek,R.: Uncertain Information Processing in Expert Systems. CRC Press: Boca Raton–Ann Arbor–London–Tokyo, 1992.

[9] Henrion, M.: Propagating uncertainty in Bayesian networks by probabilistic logic sampling. In: Lemmer, J.F.; Kanal, L.N. (eds.): Uncertainty in Artificial Intelligence 2. Elsevier Science Publishers B.V. (North–Holland), 1988.

[10] Jensen, F.V.; Lauritzen, S.L.; Oleson, K.G.: Bayesian updating in causal probabilistic networks by local computations. Computational Statistics Quaterly 4 (1990), 269-282.

[11] Kolmogorov, A.: Grundbegriffe der Wahrscheinlichkeitsrechnung. Springer: Berlin, 1933. (Reprint: Springer: Berlin–Heidelberg–New York, 1973.)

[12] Kruse, R; Schwecke, E.; Heinsohn, J.: Uncertainty and Vagueness in Knowledge Based Systems: Numerical Methods. Springer: New York–Berlin–Heidelberg–London–Paris–Tokyo, 1991.

[13] Kruse, R; Gebhardt, J.; Klawonn, F.: Fuzzy Systeme. Teubner: Stuttgart, 1993.

[14] Lauritzen, S.L.; Dawid, A.P.; Larsen, B.N.; Leimer, H.–G.: Independence properties of directed Markov fields. Networks 20 (1990), 491-505.

[15] Lauritzen, S.L.; Spiegelhalter, D.: Local Computations with Probabilities on Graphical Structures and Their Application to Expert Systems. J. Roy. Stat. Soc. B, 50 (2) (1988), 157-224.

[16] Liesenfeld, B.; Hierle, A.: A new clinical decision support tool for differential diagnosis, likelihood of development and progression of diabetic nephropathy in insulin–dependent diabetics. Proceedings of the International Conference on Neural Networks and Expert Systems in Medicine and Healthcare, Plymouth, August 24-26, 1994.

[17] Matthes, R.; Oppel, U.G.: Convergence of causal probabilistic networks. In: Bouchon–Meunier, B., Valverde, L.; Yager, R.R.(eds.): Intelligent Systems with Uncertainty. North–Holland: Amsterdam-London–New York–Tokyo, 1993.

[18] Neapolitan, R.E.: Probabilistic Reasoning in Expert Systems. Theory and Algorithms. J. Wiley and Sons: New York–Chichester–Brisbane–Toronto–Singapore, 1990.

[19] Neveu, J.: Mathematische Grundlagen der Wahrscheinlichkeitstheorie. Oldenburg: M"unchen, 1969.

[20] Oppel, U.G.: Every Complex System Can be Determined by a Causal Probabilistic Network without Cycles and Every Such Network Determines a Markov Field. In:Kruse, R.; Siegel, P. (eds.): Symbolic and Quantitative Approaches to Uncertainty. Lecture Notes of Computer Science 548. Springer: Berlin–Heidelberg–New York, 1991.

[21] Oppel, U.G.; Hierle, A.; Janke, L.; Moser, W.: Transformation of Compartmental Models into Sequences of Causal Probabilistic Networks. In: Andreassen, S.: Engelbrecht, R.; Wyatt, J. (eds.): Artificial Intelligence in Medicine. IOS Press: Amsterdam–Oxford–Washington–Tokyo, 1993.

[22] Oppel, U.G.: Causal probabilistic networks and their application to metabolic processes. To appear in: Mammitzsch, V.; Schneeweiß, H.: Proceedings of the Second Gauss Symposium, Munich, August 1993.

[23] Oppel, U.G.: Series of causal probabilistic networks induced by systems of differential equations for diagnosis and prognosis of metabolic processes. Proceedings of the Fifth International Conference on Information Processing and Management of Uncertainty in Knowledge–Based Systems (IPMU), July 4-8, 1994, Paris, France.

[24] Parthasarathy, K.R.: Probability Measures on Metric Spaces. Academic Press, New York–London, 1967.

[25] Pearl, J.: Probabilistic Reasoning in Intelligent Systems: Networks of Plausible Inference. Morgan Kaufmann: San Matteo, 1988.

[26] Richter, H.: Zur Grundlegung der Wahrscheinlichkeitstheorie. Math. Ann. 125, 129-139, 223-234, 335-343 (1953); 126, 362-374 (1953); 128, 305-339 (1954).

[27] Richter, H.: Wahrscheinlichkeitstheorie. Springer: Berlin–Heidelberg, 1956.

[28] Salzsieder, E.; Fischer, U.; Hierle, A.; Oppel, U.G.: A Causal Probabilistic Network Associated to the Karlsburg Model of the Glucose–Insulin Metabolism for Approximate Assessment of Parameters, Diagnosis and Prognosis. Proceedings of the International Conference on Neural Networks and Expert Systems in Medicine and Healthcare, Plymouth, August 24-26, 1994.

[29] Savage, L.R.: The Foundations of Statistics. J. Wiley and Sons: New York, 1954. Dover Publications: New York, 1972.

[30] Shachter, R.D.: A linear approximation method for probabilistic inference. In: Shachter, R.D.; Levitt, T.S.; Kanal, L.N.; Lemmer, J.F. (eds): Uncertainty in Artificial Intelligence. Elsevier Science Publishers B.V. (North–Holland), 1990.

[31] Shachter, R.D.; Peot, M.A.: Simulation approaches to general probabilistic inference on belief networks. In: Henrion, M.; Shachter, R.D.; Kanal, L.N.; Lemmer, J.F. (eds.): Uncertainty in Artificial Intelligence 5. Elsevier Science Publishers B.V. (North–Holland), 1990.

[32] Stegmüller, W.: Personelle und statistische Wahrscheinlichkeit. I and II. Springer: Berlin–Heidelberg–New York, 1973.

CONDITIONAL EVENTS AND PROBABILITY IN THE APPROACH TO ARTIFICIAL INTELLIGENCE THROUGH COHERENCE

R. Scozzafava
University "La Sapienza", Rome, Italy

ABSTRACT

By relying on the most general concept of event as a proposition, our approach refers to an *arbitrary* family of conditional events, which represent uncertain statements in an expert system. Then probability is interpreted as a measure of the degree of belief in a proposition in a given context, which is expressed in turn by another proposition. Usually it is not realistic to make probability evaluations for all possible envisaged conditional events: yet, requiring coherence of a function P, assessed only for a few conditional events of initial interest, entails that this P is a conditional probability. So the assignment of P can be based on the check of coherence, which amounts to the study of the compatibility of some linear systems, whose unknowns are the probabilities of the atoms generated by the given events. It is then possible to extend probability assessments, preserving coherence, by a step-by-step assignment to further events, leading in general to not necessarily unique values of the 'new' probabilities, possibly belonging to suitable closed intervals. Dealing with coherence and the relevant concepts of

conditional events and probability is less trivial than it may appear
and gives rise to some delicate and subtle problems.

1. INTRODUCTION

In Artificial Intelligence, often we are not actually able to give
reliable numerical evaluations concerning the degrees of belief of all
uncertain situations, but only of those strictly related to the problem
at hand and well known to us. So the theory of probability as proposed
by de Finetti [1] seeems particularly suitable: it differs radically
from the usual one (based on a measure-theoretic framework), which assumes that a unique probability measure is defined on the set of
'elementary events', constituting the so-called sample space. De Finetti's approach allows instead to assess your (coherent) probability for
as many or as few events as you feel able and interested, and this has
many important theoretical and applied consequences: for example, we
recall that the axiomatic counterpart of de Finetti's theory is weaker
than the traditional Kolmogoroff's approach [2]; moreover, it makes simpler and more effective the 'operational' aspects. In particular, with
reference to the treatment of uncertainty in Artificial Intelligence,
this has been discussed in many papers: see, for instance, [3], [4],
[5], [6], [7], [8], [9].

The main results contained in the present paper, which somewhat
overlaps with [9], are drawn from the joint paper [10].

2. PROBABILISTIC APPROACH THROUGH COHERENCE

It is in general maintained that assessments of probabilities require an overall design on the whole set of all possible envisaged situations, and hence that they cannot be easily achieved. This depends on

some well established 'myths', such as the (putative) need of

(i) a beforehand given "algebraic" structure (e.g. a boolean ring, a σ-algebra, etc.) on the set of all possible conditional or unconditional events which represent the envisaged situation;

(ii) an overall probability assessment on the aforementioned family of events;

(iii) suitable assumptions (such as conditional independence, maximum entropy, and so on) aiming at getting a *unique* probability value;

(iv) a frequentist interpretation of probability, that often unnecessarily restricts its domain of applicability.

Let $\mathcal{C}$ be an *arbitrary* family of conditional events and P a real function defined on $\mathcal{C}$. Given *any* finite subfamily

$$\mathcal{F} = \{E_1|H_1, \ldots, E_n|H_n\} \subseteq \mathcal{C} , \qquad (1)$$

put $P(E_i|H_i) = p_i$ for $i = 1,\ldots,n$. Then, denoting by b the *indicator function* of an event B, we consider the random quantity

$$G = \sum_{i=1}^{n} \lambda_i h_i (e_i - p_i) \qquad (2)$$

(gain corresponding to a combination of n bets of amounts $p_1\lambda_1, \ldots, p_n\lambda_n$ on $E_1|H_1, \ldots, E_n|H_n$, with arbitrary real stakes $\lambda_1, \ldots, \lambda_n$).

Denoting by H_o the union $H_1 \cup \ldots \cup H_n$ and by $G_{|H_o}$ the *restriction* of G to H_o , we have

DEFINITION - The real function $P : \mathcal{C} \longrightarrow \mathbb{R}$ is *coherent* if, for each assessment $\mathcal{P} = (p_1, \ldots, p_n)$ on a finite family $\mathcal{F} \subseteq \mathcal{C}$, with $p_i = P(E_i|H_i)$, and for every choice of $\lambda_1, \ldots, \lambda_n \in \mathbb{R}$, the possible values of the corresponding gain $G_{|H_o}$ are neither all positive nor all negative.

Given the *atoms* A_r $(r = 1,2,\ldots,m)$ generated by the 2n events $E_1,\ldots,E_n$, $H_1,\ldots,H_n$, the possible values of $G_{|H_o}$ are those corresponding to the partition of Ω into the atoms.

It has been essentially proved by B. de Finetti that if the assessment P is coherent, then P satisfies all axiomatic properties of a (finitely additive) conditional probability :

THEOREM 1 - Let $\mathcal{E}$, $\mathcal{H}$ be two *arbitrary* families of events, such that $\Omega \in \mathcal{E}$ and $\phi \notin \mathcal{H}$. If P on $\mathcal{C} \subseteq \mathcal{E} \times \mathcal{H}$ is *coherent*, then

(a) given any $H \in \mathcal{H}$ and $A_1, \ldots, A_n \in \mathcal{E}$ such that also their union belongs to $\mathcal{E}$ and $A_i A_j \subseteq H^c$ $(i \neq j)$, the function $P(\cdot | H)$ defined on $\mathcal{E}$ satisfies

$$P\left(\left(\bigcup_{k=1}^{n} A_k \right) \middle| H \right) = \sum_{k=1}^{n} P(A_k | H) \quad , \quad P(\Omega | H) = 1 \; ;$$

(b) $P(H|H) = 1$ for any $H \in \mathcal{E} \cap \mathcal{H}$;

(c) given E, H, A, such that $E \in \mathcal{E}$, $H \in \mathcal{E}$, $EH \in \mathcal{E}$, with $A \in \mathcal{H}$ and $EA \in \mathcal{H}$, then

$$P(EH|A) = P(E|A) \, P(H|EA) \; .$$

Notice that *(c)* reduces, when $A = \Omega$, to the classical *product rule* for probability.

Essentially, this *THEOREM* states that requiring *coherence* of P on $\mathcal{C}$ entails that P is the restriction on $\mathcal{C}$ of a (finitely additive) *conditional probability* : in fact the relevant axioms defining a conditional probability are just *(a)*, *(b)*, *(c)*, where $\mathcal{E}$ is a field and $\mathcal{H}$ an additive class.

We can assess instead the probabilities without referring to some restriction of an overall assignment. A *direct check of coherence* refers to an *arbitrary* family $\mathcal{C} \subseteq \mathcal{E} \times \mathcal{H}$ of conditional events, *with no underlying structure* : so *it is possible to assess P only on a set of events of interest.*

Moreover, since the assessment of conditional probabilities is bounded to satisfy only the requirement of coherence, there is no need, as in Kolmogorov's approach, where the conditional probability $P(E|H)$ is *by definition* the ratio $P(E \cap H)/P(H)$, of assuming positive probability for the conditioning event, or, in the continuous case, of knowing *the whole* conditioning distribution, as required by the usual Radon-Nikodym framework: in other words, $P(E|H)$ can be assessed on the ground of coherence *only* and makes sense for *any pair* of events E, H, with $H \neq \emptyset$.

We need also the following extension theorem (again essentially due to de Finetti):

THEOREM 2. Given an assessment P on a class $\mathscr{C}$ of conditional events and an arbitrary class $\mathscr{K} \supseteq \mathscr{C}$, then there exists a (possibly not unique) coherent extension of P to $\mathscr{K}$ if and only if P is coherent on $\mathscr{C}$.

In particular, if $\mathscr{K} = \mathscr{C} \cup \{E|H\}$ and $P(E|H) = p$, coherent assessments of the conditional probability p are those of a suitable closed interval $[p',p''] \subseteq [0,1]$, with $p' \leq p''$.

Also a clear-cut distinction between the *meaning* of probability and the various multifacet *methods of assessment* is essential. With respect to the meaning, probability can be regarded as a measure of *the degree of belief* hold by the *subject* that is making the assessment. Nevertheless, other most 'popular' and well known approaches to probability may be taken as useful methods of assessment (based on *combinatorial* arguments or on the *observed frequency*).

For a deeper discussion of these aspects, see [11], [12].

Moreover, to handle the probabilistic approach to Artificial Intelligence according to the theory sketched above, a formulation of uncertain statements in terms of *conditional* events is needed. They correspond to a 3-valued logic: so it is necessary to define appropriate logical relations and operations , extending the usual ones between standard events. There are many *pros* and *contras* concerning the 'right' choice among different possible definitions: usually they should depend on each specific context and application (for a deepening, see [13], [14]).

Anyway, real world situations make very significant assuming an 'open' framework and not a specific algebraic structure for the family of conditional events on which probability is assessed.

Moreover there are many multifacet controversial aspects including the need to avoid misunderstandings when the conditioning event is interpreted as a fit representation of a given information. For example, a careful distinction between *assumed* and *acquired* information is essential, i.e. an important issue concerns the need of interpreting the conditional probability $p = P(A|B)$ as *the probability of (A given B)*

rather than as

(the probability of A) given B.

The latter interpretation is unsustainable, since it would literally mean *'if B occurs, then p is the probability of* A', which is actually a form of logical deduction leading to absurd conclusions. Consider in fact a set of five balls $\{1,2,3,4,5\}$ and the probability that a number drawn from it at random is even (which is 2/5) : this probability could instead be assessed equal to 1/3, since this is the value of the sought probability *conditionally on the occurrence of each one of the events*

$$E_1 = \{1,2,3\} \quad \text{or} \quad E_2 = \{3,4,5\},$$

and one (possibly both) of them *will certainly occur.*

The content of this Section should have made it clear that dealing with coherence and the relevant concepts of conditional events and probability is less trivial than it may appear, and gives rise to some delicate and subtle problems.

3. CHECKING COHERENCE

To illustrate in more detail the concept of coherence, consider an assessment $\mathcal{P} = (p_1, \ldots, p_n)$ on an arbitrary finite family $\mathcal{F} = \{E_1, \ldots, E_n\}$ (i.e. $P(E_i) = p_i$, $i = 1,2,\ldots n$) of (unconditional) events, and denote by $A_1, A_2, \ldots, A_m$ the atoms generated by these events. Then, referring to the unknowns $x_r = P(A_r)$, coherence of $\mathcal{P}$ amounts to the compatibility of the following system

$$
\begin{cases}
\sum_{\substack{r \\ A_r \subseteq E_i}} x_r = p_i, & i = 1,\ldots,n, \\
\\
\sum_{r=1}^{m} x_r = 1, \quad x_r \geq 0, \quad r = 1,\ldots,m.
\end{cases}
\tag{$\mathcal{S}$}
$$

So coherence is equivalent to the existence of an extension of P from the given events E_i ($i = 1, 2, \ldots, n$) to the atoms generated by them.

Moreover, given a further event E_{n+1} and the corresponding extended family $\mathcal{K} = \mathcal{F} \cup \{E_{n+1}\}$, consider the case in which E_{n+1} is union of some atoms, i.e. E_{n+1} is *logically dependent* on the events of $\mathcal{F}$: then, putting $P(E_{n+1}) = p_{n+1}$, one has

$$p_{n+1} = \sum_r x_r .$$
$$A_r \subseteq E_{n+1}$$

If the vector $(x_1, x_2, \ldots, x_m)$ varies in the set of solutions of $(\mathcal{S})$, the probability p_{n+1} describes an interval $[p', p''] \subseteq [0,1]$.

A similar conclusion can be suitably reached if E_{n+1} is not logically dependent on the events of $\mathcal{F}$. This result is *the fundamental theorem of probabilities* of de Finetti. The values p', p'' can be determined by the simplex method of linear programming [15].

Many interesting and unaspected features come to the fore when one tries to extend the above theory to *conditional* events and to the ensuing and the relevant concept of *conditional probability*, once we refer to the suitable extension of the concept of coherence, as previously defined.

Let $\mathcal{F} = \{E_1|H_1, \ldots, E_n|H_n\}$ be an arbitrary finite family of conditional events and let P , with $P(E_i|H_i) = p_i$ ($i = 1, \ldots, n$), be a given coherent assessment on $\mathcal{F}$. Denote by the same symbol P a coherent extension of P onto $\mathcal{K} = \mathcal{F} \cup \{E_1, \ldots, E_n, H_1, \ldots, H_n\} \cup \mathcal{A}$, where $\mathcal{A} = \{A_1, \ldots, A_m\}$ is the set of atoms generated by the events $E_1, \ldots, E_n, H_1, \ldots, H_n$: clearly

$$\sum_{r=1}^{m} P(A_r) = 1 . \tag{3}$$

Moreover, since each event E_i or H_i is a union of atoms, we have, for example, the representation

$$P(H_i) = \sum_r P(A_r) . \tag{4}$$
$$A_r \subseteq H_i$$

Notice that (3) cannot ensure that the sum (4) is strictly positive *for every conditioning event* H_i : in this case also a representation for each conditional probability $P(E_i | H_i) = p_i$ would easily follow, i.e.

$$p_i = \frac{\sum_{r \atop A_r \subseteq E_i H_i} P(A_r)}{\sum_{r \atop A_r \subseteq H_i} P(A_r)} . \qquad (5)$$

It should be clear that if we consider an assessment $p_i = P(E_i | H_i)$ (for $i = 1, \ldots, n$) on the set $\mathcal{F}$ of conditional events $E_i | H_i$, then the existence of an extension of the probability distribution P on the set $\mathcal{A}$, satisfying the system

$$\begin{cases} P(A_r) \geq 0 , \quad r = 1, \ldots, m \\[2mm] \sum_{r \atop A_r \subseteq E_i H_i} P(A_r) = P(E_i | H_i) \sum_{r \atop A_r \subseteq H_i} P(A_r) , \quad i = 1, \ldots, n \\[2mm] \sum_{r \atop A_r \subseteq H_o} P(A_r) = 1 , \end{cases} \qquad (6)$$

is necessary, but not sufficient, to ensure coherence of P.

Notice that the second line of (6) corresponds to the *product rule*

$$P(EH) = P(E | H) \, P(H) \qquad (7)$$

for conditional probability, which is trivially satisfied by *any* value of $P(E | H)$ when $P(H) = 0$.

The third line of (6) is an easy consequence of the following

LEMMA (for the proof, see [10]) - If P is a coherent assessment on a finite family of conditional events

$$\mathcal{F} = \{E_1 | H_1, \ldots, E_n | H_n\} ,$$

then also any extension of P to $\mathcal{F} \cup \{H_o\}$ such that $P(H_o) = 1$, where H_o is the union of the H_i's, is coherent.

EXAMPLE - Consider the events $E | H$, $E^c | H$, $E | H^c$ and the assessment

$$P(E|H) = P(E^c|H) = \frac{1}{3} \ , \quad P(E|H^c) = 1 \ ,$$

which is not coherent (since property *(a)* of *THEOREM 1* does not hold). Nevertheless the probability distribution

$$P(A_1) = P(A_2) = P(A_3) = 0 \ , \quad P(A_4) = 1$$

on the four atoms $A_1 = EH$, $A_2 = E^cH$, $A_3 = E^cH^c$, $A_4 = EH^c$ satisfies (6): therefore it is not sufficient to ensure coherence.

Condition (6) becomes sufficient, allowing the representation (5), if also the following (not necessary) condition holds:

$$\sum_{\substack{r \\ A_r \subseteq H_i}} P(A_r) > 0 \quad \text{for every conditioning event } H_i . \tag{8}$$

So testing coherence of $P(E_i|H_i)$ would be equivalent to test the solvability of the linear system (6), with unknowns $P(A_r)$, under the condition (8). But, even if the latter may seem an 'almost natural' condition (at least on a finite set of events), it introduces some computational complications, due to auxiliary unknowns involving strict inequalities. This would be particularly awkward also for the mathematical programming methods needed to construct extensions of P to 'new' conditional events. Anyway, assuming (8) is not suitable in the case that the finite family $\mathcal{F}$ is a subfamily of an infinite set of conditional events.

A way out, giving up any positivity hypotheses, is possible through the following procedure: if P is coherent, then, given $\mathcal{F}$, there exists at least an extension P_o of P on $\mathcal{A}_o$ satisfying system (6), with P_o in place of P. Let H_o^1 be the union of the H_i's such that

$$\sum_{\substack{r \\ A_r \subseteq H_i}} P_o(A_r) = 0 \ ;$$

then the representation

$$P(E_i|H_i) = \frac{\displaystyle\sum_{\substack{r \\ A_r \subseteq E_iH_i}} P_o(A_r)}{\displaystyle\sum_{\substack{r \\ A_r \subseteq H_i}} P_o(A_r)}$$

holds for any $E_i|H_i$ such that H_i is not contained in H_o^1 (in particular, for *all* $E_i|H_i$ when $H_o^1 = \varnothing$). Then, denote by $\mathcal{A}_1$ the set of atoms contained in H_o^1 : coherence of P on $\mathcal{C}$ implies coherence of its restriction P_1 on the subset

$$\mathcal{F}_1 = \left\{ E_i|H_i : H_i \subseteq H_o^1 \right\} ,$$

and so there exists at least an extension of P_1 on $\mathcal{A}_1$ satisfying system (6) with P_1 in place of P and H_o^1 in place of H_o. Let H_o^2 be the union of the H_i's such that

$$\sum_r \ P_1(A_r) = 0$$
$$A_r \subseteq H_i$$

and denote by $\mathcal{A}_2$ the set of atoms contained in H_o^2 . Then for any $E_i|H_i$ such that H_i is not contained in H_o^2 the representation

$$P(E_i|H_i) = \frac{\displaystyle\sum_r \ P_1(A_r) \atop A_r \subseteq E_i H_i}{\displaystyle\sum_r \ P_1(A_r) \atop A_r \subseteq H_i}$$

holds. And so on.

We are able in fact to establish the following theorem (see [6], [10]), which gives a characterization of coherent conditional probabilities in terms of classes of probability distributions defined on suitable subsets of the set $\mathcal{A}$ of atoms generated by the given conditional events.

THEOREM 3 - Let $\mathcal{C}$ be an arbitrary family of conditional events. For a real function P on $\mathcal{C}$ the following two statements are equivalent:

(i) P is a coherent conditional probability on $\mathcal{C}$;

(ii) for any finite subset $\mathcal{F} = \{E_1|H_1,\ldots,E_n|H_n\}$ of $\mathcal{C}$, denoting by $\mathcal{A}_o$ the relevant set of atoms, there exist classes of probabilities $\{P_\alpha\}$ defined on suitable subsets $\mathcal{A}_\alpha \subseteq \mathcal{A}_o$, with $\mathcal{A}_{\alpha'} \subset \mathcal{A}_{\alpha''}$ for $\alpha' > \alpha''$ and $P_{\alpha''}(A_r) = 0$ if $A_r \in \mathcal{A}_{\alpha'} \subset \mathcal{A}_{\alpha''}$, such that for any $E_i|H_i \in \mathcal{F}$ there is (for each class) a unique P_α with

$$\sum_r \; P_\alpha (A_r) > 0 \quad \text{and} \quad P(E_i | H_i) = \frac{\displaystyle\sum_{\substack{r \\ A_r \subseteq E_i H_i}} P_\alpha (A_r)}{\displaystyle\sum_{\substack{r \\ A_r \subseteq H_i}} P_\alpha (A_r)} \; . \tag{9}$$

In conclusion, we can proceed along the following steps:

1) given a set $\mathcal{F}$ of n conditional events $E_1 | H_1, \ldots, E_n | H_n$, supply all the known logical relations among the events $E_1, \ldots, E_n$, $H_1, \ldots, H_n$, so that the cardinality m of the relevant set $\mathcal{A}_o$ of atoms will be usually much less than 2^{2n}.

2) given the assessment

$$p_1 = P(E_1 | H_1) \; , \ldots, \; p_n = P(E_n | H_n) \; , \tag{*}$$

introduce the system, with unknowns $P_\alpha (A_r)$,

$$
\begin{cases}
P_\alpha (A_r) \ge 0 \; , \quad A_r \in \mathcal{A}_\alpha \\[2ex]
\displaystyle\sum_{\substack{r \\ A_r \subseteq E_i H_i}} P_\alpha (A_r) = P(E_i | H_i) \displaystyle\sum_{\substack{r \\ A_r \subseteq H_i}} P_\alpha (A_r) \; , \quad \left[\text{if } P_{\alpha - 1} (H_i) = 0 \right] \\[2ex]
\displaystyle\sum_{\substack{r \\ A_r \subseteq H_o^\alpha}} P_\alpha (A_r) = 1
\end{cases}
\tag{**}
$$

where $P_{-1} (H_i) = 0$ for all H_i's, and H_o^α denotes, for $\alpha \ge 0$, the union of the H_i's such that $P_{\alpha - 1} (H_i) = 0$; so, in particular,

$$H_o^o = H_o = H_1 \cup \ldots \cup H_n \; .$$

3) Put $\alpha = 0$ in (**).

4) if (**) has no solutions, the assessment (*) is *not* coherent and must be revised.

5) if (**) has a solution $P_\alpha (A_r)$ such that

$$P_\alpha (H_i) = \sum_{\substack{r \\ A_r \subseteq H_i}} P_\alpha (A_r) > 0 \tag{***}$$

for every H_i specified in the second line of (**), then the assessment

(*) is coherent, while if it has *only* solutions such that

$$P_\alpha(H_i) = 0 \text{ for some } H_i,$$

proceed as follows.

6) put $\alpha + 1$ in place of α, and go to steps 4) and 5) until the exhaustion of the H_i's.

If in step 5) we find *both kinds of solutions*, the assessment is obviously coherent, but we can determine all the classes $\{P_\alpha\}$ by going on to the following step

6') put $\alpha + 1$ in place of α, and seek solutions of (**) satisfying (***), until the exhaustion of the H_i's.

This optional step is necessary to get coherent extension to new conditional events.

4. EXTENSIONS OF PROBABILITY ASSESSMENTS

Given a new conditional event $E|H$, there exists a suitable closed interval $[p_*, p^*] \subseteq [0,1]$ such that an extension p of P from the finite subset

$$\mathcal{F} = \{E_1|H_1, \ldots, E_n|H_n\}$$

to $E|H$ is coherent if and only if it satisfies

$$p_* \leq p_{n+1} \leq p^* .$$

We proved in [10] that these two bounds can be characterized (as in the case of unconditional events) as probabilities $P(E_*|H_*)$ and $P(E^*|H^*)$ of suitable conditional events that are *logically dependent* on $\mathcal{F}$, i.e. such that the events E_*, H_*, E^*, H^* are union of atoms. In particular, $E_*|H_*$ and $E^*|H^*$ are, respectively, the 'maximum' and the 'minimum' conditional event logically dependent on $\mathcal{F}$ satisfying

$$E_*|H_* \subseteq_\circ E_{n+1}|H_{n+1} \subseteq_\circ E^*|H^* ,$$

where the inclusion $\subseteq_\circ$ between conditional events is defined as in Goodman & Nguyen [16]:

$$A|H \subseteq_\circ B|K \iff AH \subseteq BK \text{ and } B^cK \subseteq A^cH .$$

Moreover, for $A|H$, $B|K$ belonging to an *arbitrary* set $\mathcal{C}$ of conditional events and for a *coherent* P on $\mathcal{C}$, it has been proved in [5] that

$$A|H \subseteq_\circ B|K \implies P(A|H) \le P(B|K) .$$

THEOREM 4 - Let $\mathcal{F}$ be a finite family of conditional events and let P be coherent on $\mathcal{F}$. Given a further event $E|H$, denote by $E_*|H_*$ $E^*|H^*$ the 'maximum' ['minimum'] event logically dependent on $\mathcal{F}$ contained in $E|H$ [containing $E|H$]. Then a coherent assessment of the conditional probability $P(E|H)$ is any value in the closed interval

$$p_* \le P(E|H) \le p^* ,$$

with

$$p_* = \inf_{\{p_\alpha^{H_*}\}} P(E_*|H_*) \quad , \quad p^* = \sup_{\{p_\alpha^{H^*}\}} P(E^*|H^*) ,$$

if $\{p_\alpha^{H_*}\} \ne \varnothing$ and $\{p_\alpha^{H^*}\} \ne \varnothing$, where $\{p_\alpha^H\}$ denotes all the subclasses of the classes $\{P_\alpha\}$ (defined in *THEOREM 3*) whose families contain a probability P_γ such that $P_\gamma(H) > 0$ (γ depends on the family), otherwise $p_* = 0$ or $p^* = 1$.

It is clear that, for a coherent assessment of the probability of a new conditional event $E|H$, it is enough to discuss only the case of $E|H$ *logically dependent* on the family $\mathcal{F}$.

By step 6') of the previous algorithm, we are able to determine all the classes $\{P_\alpha\}$: if $P_\alpha(H) = 0$ for all α such that $H \subseteq \mathcal{A}_\alpha$ and for any class $\{P_\alpha\}$, then *any* value between 0 and 1 can be assigned to the probability $p = P(E|H)$ of the new event $E|H$, otherwise we *should* solve the problem of *fractional* programming

$$p_* = \inf_{\{p_\alpha^H\}} \frac{\sum_r P_\gamma(A_r)}{\sum_r P_\gamma(A_r)} \quad , \quad p^* = \sup_{\{p_\alpha^H\}} \frac{\sum_r P_\gamma(A_r)}{\sum_r P_\gamma(A_r)} ,$$

with the numerator sums over $A_r \subseteq EH$ and the denominator sums over $A_r \subseteq H$.

where γ has the meaning just explained and the $P_\gamma(A_r)$'s are solutions of the systems

$$
\begin{cases}
P_\gamma(A_r) \geq 0 , \quad A_r \in \mathscr{A}_\gamma \\[2ex]
\sum_{\substack{r \\ A_r \subseteq E_i H_i}} P_\gamma(A_r) = P(E_i \mid H_i) \sum_{\substack{r \\ A_r \subseteq H_i}} P_\gamma(A_r) , \quad \left[\text{if } P_{\gamma-1}(H_i) = 0 \right] \\[3ex]
\sum_{\substack{r \\ A_r \subseteq H_o^\gamma}} P_\gamma(A_r) = 1
\end{cases}
$$

This problem can be suitably transformed into one of *linear* programming (see [10]).

5. CONCLUSIONS

Usually it is very difficult to build expert systems which have a sufficient degree of flexibility with respect to the relevant probability judgments. It is in general maintained that assessments of probabilities require an overall design on the whole set of all possible envisaged situations, and hence that they cannot be easily achieved. A problem that often occurs in Artificial Intelligence is the following: the field expert is actually able to give reliable numerical evaluations of the degree of belief concerning only a few uncertain situations, i.e. the ones strictly related to the problem at hand and known by him.

Aim of this paper has been to exhibit the usefulness of the probabilistic approach based on de Finetti's theory of coherence to deal with this kind of partial and vague information by a step-by-step assignment of the relevant probabilities.

REFERENCES

1. De Finetti, B.: Teoria delle probabilita`, Einaudi, Torino 1970 (English transl.: Theory of Probability, Wiley, Chichester 1974).

2. Scozzafava, R.: Probabilita` condizionate: de Finetti o Kolmogorov?, in: Scritti in omaggio a L. Daboni, LINT, Trieste 1990, 223-237.

3. Gilio, A. and R. Scozzafava: Le probabilita` condizionate coerenti nei sistemi esperti, in: Atti Giornate A.I.R.O. su 'Ricerca Operativa e Intelligenza Artificiale', IBM, Pisa 1988, 317-330.

4. Coletti, G., Gilio, A. and R. Scozzafava: Conditional events with vague information in expert systems, in: Lecture Notes in Computer Sciences n.521 (Eds. B. Bouchon-Meunier, R.R. Yager, and L.A. Zadeh), Springer-Verlag, Berlin 1991, 106-114.

5. Coletti, G., Gilio, A. and R. Scozzafava: Comparative probability for conditional events : a new look through coherence, Theory and Decision, 35 (1993), 237-258.

6. Coletti, G.: Numerical and qualitative judgments in probabilistic expert systems, in: Proc. Intern. Workshop on 'Probabilistic Methods in Expert Systems' (Ed. R. Scozzafava), SIS, Roma 1993, 37-55.

7. Coletti, G.: Coherent Numerical and Ordinal Probabilistic Assessments, IEEE Transactions on Systems, Man, and Cybernetics, 24 (1994), in press.

8. Coletti, G. and R. Scozzafava: A Coherent Qualitative Bayes' Theorem and its Application in Artificial Intelligence, in: Proc. 2nd Intern. Symp. on 'Uncertainty Modeling and Analysis', Maryland (Ed. B.M. Ayyub), IEEE Computer Society Press, Los Alamitos 1993, 40-44.

9. Scozzafava, R.: Probabilistic background for the management of uncertainty in Artificial Intelligence, European Journal of Engineering Education, in press.

10. Coletti, G. and R. Scozzafava: Characterization of coherent conditional probabilities, working paper (submitted).

11. Scozzafava, R.: A merged approach to Stochastics in Engineering Curricula, European Journal of Engineering Education, 15, n.3 (1990),

243-250.

12. Scozzafava, R.: Probability assessment, in: Proc. Intern. School of Mathematics G.Stampacchia on 'Probability and Bayesian Statistics in Medicine and Biology', Erice (Eds. I. Barrai, G. Coletti, M. Di Bacco), CNR Applied Math. Monographs n.4, 1991, 244-255.

13. Gilio, A. and R. Scozzafava: Conditional events in probability assessment and revision, IEEE Transactions on Systems, Man, and Cybernetics, 24 (1994), in press.

14. Goodman, I.R., Nguyen, H.T. and E.A. Walker: Conditional Inference and Logic for Intelligent Systems: a theory of measure-free conditioning, North-Holland, Amsterdam 1991.

15. Bruno, G. and A. Gilio: Applicazione del metodo del simplesso al teorema fondamentale per le probabilita' nella concezione soggettivistica, Statistica, 40 (1980), 337-344.

16. Goodman, I.R. and H.T. Nguyen: Conditional objects and the modeling of uncertainties, in: Fuzzy Computing (Eds. M.Gupta and T.Yamakawa), North Holland, Amsterdam 1988, 119-138.

ABDUCTION AND INDUCTION BASED ON NON-MONOTONIC REASONING

F. Bergadano

University of Messina, Messina, Italy

and

Ph. Besnard

IRISA, Rennes, France

Abstract. We consider abduction and induction in Artificial Intelligence, focussing on one particular case for each, namely diagnostic and learning. Overall, we formalize abduction as well as induction as two syntactic specializations of a single reasoning scheme, leading from observed consequences to plausible hypotheses. The problem of finding hypotheses that justify given facts is then transformed into an inference, leading from these very facts, and from prior relevant background knowledge, to the corresponding hypotheses. This is found to be actually a form of non-monotonic inference, amenable to some appropriate non-monotonic logic, where the direction of reasoning is reversed: A logic of "reversing implication" is obtained. We give a sequent system for that logic, with application to both an example of abduction and an example of induction.

1 Introduction

We start by claiming that abduction and induction (which are both to be described in a moment) are useful forms of reasoning. Even though they fail to be correct forms of reasoning (see below), they are effective to the point that their usage has been carefully studied and advocated [15] in the domain ruled by the most stringent account of reasoning, namely mathematics.

Here is first an example [15] among many others of a case involving abduction. The story is that Euler got a proof of the following statement:

If for any integer n,
$$8n + 3 = q^2 + 2p \text{ for some integer } q \text{ and some prime } p \qquad (1)$$

then for any integer n, there exist x, y, and z such that

$$n = \frac{x(x-1)}{2} + \frac{y(y-1)}{2} + \frac{z(z-1)}{2} \qquad (2)$$

Considering the latter property (2), in view of the whole statement, Euler was led to the following hypothesis: Any integer of the form $8n + 3$ is the sum of a square and of the double of a prime.

The pattern is that, from (2) and from the fact that (1) implies (2), it is worth considering (1).

This is just *abduction*: From B and from A implies B, conclude A (tentatively). Importantly, A is of course drawn only on condition that it is not known to be false (in a strict sense. abduction is not a correct form of reasoning).

Let us turn to an illustration [14] for induction. Investigating the decomposition of natural numbers in sums of squares, one comes up with observations like these:

$$1 = 1^2$$
$$2 = 1^2 + 1^2$$
$$3 = 1^2 + 1^2 + 1^2$$
$$4 = 2^2$$
$$5 = 1^2 + 2^2$$
$$6 = 1^2 + 1^2 + 2^2$$
$$7 = 1^2 + 1^2 + 1^2 + 2^2$$
$$8 = 2^2 + 2^2$$
$$9 = 3^2$$
$$\vdots$$
$$31 = 1^2 + 1^2 + 2^2 + 5^2$$
$$32 = 4^2 + 4^2$$
$$33 = 1^2 + 4^2 + 4^2$$
$$34 = 3^2 + 5^2$$
$$35 = 1^2 + 3^2 + 5^2$$
$$\vdots$$

Now, all natural numbers that have been examined above turn out to be the sum of at most four squares. At this point, the statement that any natural number is the sum of at most four squares is worth considering.

Abbreviating "n is the sum of at most four squares" by $S(n)$, all this means that from $S(1), \ldots, S(9), S(31), \ldots, S(35)$ then $\forall n S(n)$ is concluded (tentatively).

This is *induction*: If there is some m and some statement A that is verified for m cases, $A_1, \ldots, A_m$, then $\forall x A(x)$ arises as a tentative conclusion. Similarly to abduction, the conclusion is drawn only on condition that it is not known to be false (similarly as well, induction is not a correct form of reasoning in a strict sense).

Part of the former example can also be used to give an illustration for induction, maybe a little bit less convincing because the reader may doubt that anybody would spontaneously observe the following:

$$11 = 1 + 2 \times 5$$
$$19 = 9 + 2 \times 5$$
$$27 = 1 + 2 \times 13$$
$$35 = 1 + 2 \times 17$$
$$43 = 9 + 2 \times 17$$
$$51 = 25 + 2 \times 13$$
$$59 = 49 + 2 \times 5$$
$$\vdots$$

but if somebody does, then that person surely considers adopting $8n + 3 = q^2 + 2p$ as an hypothesis (of course, $8n + 3 = q^2 + 2p$ is simply a shorthand for (1)).

Finally, it is important to realize that both abduction and induction take place in the context of some background theory (in the above examples, number theory). That is, the reasoning patterns just described must be understood as working with some of their premises (and additional ones) coming from the background theory.

2   Logical forms of abduction and induction

We intend to consider, in a logical setting, abduction and induction as applied in Artificial Intelligence. Under such a view, the following two fundamental features must be pointed out.

As regards abduction, it appears that the premises of the abductive reasoning scheme do not conform to deductive reasoning (we take first order logic as a formal counterpart to deductive reasoning). Should it be the case, the premise "B" would then subsume the premise "A implies B" (in the sense that the truth of a given statement B entails the truth of any statement of the form "A implies B"). Abduction would thus amount to concluding A from B, which is nonsense.

As regards induction, it must now be said that the amount of evidence needed for induction, which should be rather large in mathematical reasoning as in the above examples, can in certain ways (related to practical matters) reduce to a single premise when it comes to machine learning.

For reference below, let us remind the reader that an inference scheme is written

$$\frac{P_1, P_2, \ldots, P_n}{C}$$

and that it reads: From all of $P_1, P_2, \ldots, P_n$ (the premises), infer C (the conclusion).

We are now in position to discuss abduction and induction as applied in Artificial Intelligence, where they are considered distinct, so that they have generated separate fields of study. After a first and simple analysis, one finds in effect distinct patterns. For abduction:

$$\frac{\forall x\ (P(x) \to Q(x)), Q(a)}{P(a)}$$

For induction:

$$\frac{P(a)}{\forall x P(x)}$$

Deeper analysis, however, suggests that the difference between the two schemes is not always easy to state. E.g., using the tautology $\Phi = (\forall x P(x)) \to P(a)$ one gets:

$$\frac{\Phi, P(a)}{\forall x P(x)}$$

as an abductive inference step, but tautologies being taken as granted, it actually has the same premise and conclusion as the inductive inference scheme. It would then seem that the abductive scheme includes simple forms of inductive reasoning. However, the conclusions of abduction are usually specific facts (e.g., my car engine is broken), whereas induction yields general rules, with universally quantified variables (e.g., all men are mortal). Yet, we here disregard such a level of distinction.

In fact, we are interested in formalizing both abduction and induction as syntactic specializations of a single reasoning scheme, leading from observed consequences to plausible hypotheses. The problem of finding hypotheses that justify given facts is then transformed into an *inference*, leading from these very facts, and from relevant background knowledge, to the corresponding hypotheses. This is found to be a form of non-monotonic inference, amenable to a special non-monotonic logic, where the direction of reasoning is reversed: A logic of "reversing implication" is obtained.

3   From observations to hypotheses: a logic-based analysis

We assume familiarity with first order logic [4], as we will use words such as ground atoms and literals without defining them.

Regarding notation, the symbols $\neg, \vee, \wedge, \rightarrow$ denote as usual the connectives for negation, disjunction, conjunction, and implication. The symbol $\leftarrow$ will also be used at a few places throughout the text to denote converse implication, that is, $A \leftarrow B$ is equivalent to $B \rightarrow A$, intuitively meaning "if B then A". Also, the existential quantifier is denoted by $\exists$ ("there exists") whereas the universal quantifier is denoted by $\forall$ ("for all").

Last, $\vdash$ denotes deduction in first order logic. Notice that, whenever S and T are sets of formulas, $T \vdash S$ abbreviates $T \vdash s$ for all s in S whereas $T \nvdash S$ abbreviates $T \nvdash s$ for all s in S, i.e. there is no s in S such that $T \vdash s$.

Given any language for first order logic, we will consider three sets of formulas, as follows:

M = Observed Manifestations.

H = A Posteriori Hypotheses.

T = A Priori Theory.

This naturally amounts to a formal, logic-based, counterpart to the entities that we have shown in the previous sections to occur in both abductive reasoning and inductive reasoning. The above three sets may be required to meet some particular syntactic requirements; for instance, the observed manifestations may be restricted to ground atoms.

There remains to give a logic-based account of the relationships to be found between those entities when abduction and induction take place. The general idea is that the following relation must hold:

$$T \wedge H \vdash M \tag{3}$$

where T and M are given whereas H must be found by some reasoning mechanisms. Moreover, the following consistency condition must hold:

$$T \wedge H \nvdash M' \tag{4}$$

where M' may be defined in a number of different ways.

Clearly, the consistency condition takes care of all hypotheses which are known to be false. The simplest case (one which is always in force) corresponds to M' being the logical contradiction, stating that no hypothesis is acceptable that contradicts the background theory T. In other words, we rule out the possibility that observed manifestations lead to some revision [7] of the background theory: The point here is only about abduction and induction, not about topics such as theory development and the like.

We shall show next that abduction and induction, as understood in recent work in Artificial Intelligence about diagnostic and learning, fall under the scheme defined by (3) and (4), with some significant differences.

4 Abduction

In accordance with the discussion so far, it appears that, in abduction, the rôle of the theory T is fundamental in the following sense:

$$H \not\vdash M \qquad (5)$$

i.e., an abductive hypothesis is not a generalization of the manifestations, but rather a fact that the theory T makes it to imply the manifestations M.

Conjoining (3), (4) and (5), we now have the logic-based *abductive scheme* that we were aiming at.

Before turning to the additional conditions on M, M', H and T that characterize abduction in contrast to induction, we ought first to make the reader aware of the following technical detail: Among ground literals (including ground atoms), one must distinguish between observable ones and non-observable ones. As for an illustration, a broken leg may be observable (so that one would decide to classify the ground atom $broken(leg)$ for instance as an observable literal), but a broken heart may not (hence one could classify $broken(heart)$ for instance as a non-observable literal).

Then, the following general conditions are normally in use that impose M, M', H and T to be:

- M: A set of ground formulas over observable literals. The elements of M are here called *observations*.
- H: A conjunction of ground disjunctions over non-observable atoms, describing a set of alternative *causes* of the observations.
- T: An arbitrary theory (no restriction).
- M': The empty clause, i.e. the absurd formula.

Typically, the theory T may contain a description of the functioning principles of some machine (including a formula such as $thick\_oil \rightarrow ringing(bell_1)$ for instance). Accordingly, the observations M may describe the observed behaviour of the machine (e.g., if two bells sounds the same and are next to each other, one may hear a bell ringing without being able to tell which one actually does; then, one could have $ringing(bell_1) \lor ringing(bell_2)$ for instance in M). Finally, the hypothesis H may be some diagnosis (e.g., $broken(engine) \land thick\_oil$).

All this (occasionally at the cost of a few changes in the above general conditions) covers ideas such as critical reasoning [16] and can as well account for a representation by models [9] used to find minimal explanations to some observation. Also relevant but less formalized is case-based explanation [11]. Moreover, abduction with explicit uncertainty [8] can be captured when the above framework is enriched with an appropriate model for uncertainty.

Last, we urge the reader to observe that the premises B and $A \rightarrow B$ of abduction as presented in the introduction, are kept apart because the former is an element of M and the latter is in T. Accordingly, the two premises are not combined by deductive reasoning and the dreadful subsumption mentioned at the beginning of Section 2 does not happen.

5 Induction

We proceed for induction as we just did for abduction, that is, we supplement (3) and (4) with an appropriate third property which is to be typical of inductive reasoning. Such an item is simply to come from the idea that inductive reasoning in the context of a background theory does not coincide with mere deduction from that theory. Stated otherwise, induction is such that the hypothesis H must actually contribute to M being true, i.e. one must have

$$T \nvdash M. \tag{6}$$

so that the hypothesis really goes beyond the knowledge already available in the form of the background theory.

Conjoining (3), (4) and (6), we now have the logic-based *inductive scheme* that we were aiming at.

Besides, the following general conditions are normally in use that impose M, M', H and T to be:

- M: A set of ground atoms. The elements of M are here called *positive examples*.
- H: A conjunction of formulas of the type

$$(\forall x_1, \ldots, x_n)\ P(x_1, \ldots, x_n) \leftarrow \Phi$$

  where Φ is usually existential and conjunctive. Moreover, the predicate P is normally required to occur in M and M'. The predicate P is called a *concept* and should be the predicate symbol in the positive examples as well as in the negative examples (see below).
- T: An arbitrary theory (no restriction).
- M': A set of ground atoms. The consistency condition (4) rules out hypotheses that are too general and would entail undesirable statements. The elements of M' are here called *negative examples*.

Variations of these four conditions and sometimes of the whole framework usefully describe recent machine learning keywords:

Multiple Predicate Learning. If the definitions in H are extended so as to be for more than one predicate P, then the so-called multiple predicate learning takes place [1, 3]. For instance, rules can then be learned for computing both sum and product, given examples of both, and a definition of successor in the background theory T.

Predicate Invention. If the predicate P occurring in H as above does not occur in the examples M and is not completely defined in the theory T, the learning algorithm that outputs H given T and M is said to perform predicate invention [13]. For instance, one could learn rules for computing sum and product, given examples of product only, and without having sum defined in T. A related notion, although less formalized, is that of *constructive* learning [12].

Approximately Correct Hypotheses. Given now a maximum observed error ϵ, we require (3) and (4) to hold for most examples, i.e. for $|M \cup M'|(1 - \epsilon)$ of them. This is especially used for predictions about future performance.

6 The logic of reversing implication

As already mentioned before, we would like to transform the abductive/inductive problem of finding H such that

$$T \wedge H \vdash M \tag{7}$$

into an *inference*

$$T \wedge M \leadsto H \tag{8}$$

where H will play the rôle of some sort of conclusion, given the background theory T and the manifestations M.

A first comment to be made is that the desired inference scheme for (8) would have to be non-monotonic (strictly speaking, $\leadsto$ being monotonic would consist of $U \leadsto V$ necessarily entailing $U \wedge W \leadsto V$). Indeed,

need not imply
$$T \wedge M \leadsto H$$
$$T \wedge M \wedge A \leadsto H.$$

This is easily seen in inductive reasoning, where A would be a new example, which might invalidate previously formed hypotheses (for instance, it might be the case that $\neg A$ is an acceptable hypothesis in view of T and M, that is, $T \wedge M \leadsto \neg A$ but of course $T \wedge M \wedge A \not\leadsto \neg A$ because no hypothesis is acceptable that contradicts some example). More generally, $T \wedge M \wedge A \leadsto H$ is refuted whenever $T \vdash A \to \neg H$.

There is even more to it than simply non-monotonicity. The point is that there exist non-monotonic inference schemes which preserve logical equivalence as induced by first order logic: Whenever U and W are logically equivalent (in symbols, $U \vdash W$ and $W \vdash U$), any formula F follows from U iff F follows from W where "follows" stands for such a non-monotonic inference scheme. This is not the case here. It is easy to find T, T', M and H such that $T' \vdash T \wedge M$ and $T \wedge M \vdash T'$ but $T \wedge M \not\leadsto H$ whereas $T' \leadsto H$. For instance, taking T to be $A \to B$ and M to be B, then $T \wedge M \leadsto A$ is expected but taking T' to consist of $C \to B$ and B, it is obvious that $T' \leadsto A$ is undesirable.

Second, going from (7) to (8) we have reversed the direction of implication. Indeed, we can consider reasoning within the background theory T. So, going from (7) to (8) amounts to moving from the first order logic deduction of H implying M to the non-classical inference of M implying H where in each case the inference takes the theory T into account as being part of the premises. It is then natural to think of some "logic of reversing implication" for performing inference (8), e.g. reminiscent of the so-called predicate completion [2]. If one uses conditional knowledge in T with all implications reversed, one may be able, after proper formalization, to turn every deduction step of the form (7) an inference step of the form (8).

Third, we would like to chose some adequate logic of "reversing implication", noticing that the inductive/abductive problem (7) is usually underdetermined and has multiple solutions. As a consequence, some preference relation over the possible hypotheses is usually adopted. In induction, this is called a *bias* and reflects prior knowledge about the future status of particular hypotheses. In abduction, a special case is that of selecting which predicates are *abducible*, i.e. which are admitted to occur in H. So, an appropriate logic could be based on some extended notion of preferential models, not the one usually adopted for non-monotonic systems [10], but one whose model theory incorporates some prior knowledge about hypotheses.

We present now a sequent system [6] for "reversing implication", considering formulas as defined in the usual way.

The *sequents* are of the form $\Gamma \leadsto \Theta$ where Γ and Θ are finite sets of formulas (i.e., the so-called permutation and merge, that allow us to display sets as lists with possibly multiple occurrences, are rules of the system).

Intuitively, a sequent $P_1, P_2, \ldots P_n \leadsto C_1, C_2, \ldots, C_m$ means that at least one of $C_1, C_2, \ldots, C_m$ follows from all of $P_1, P_2, \ldots, P_n$. The sequent system consists of rules to derive inferences, in the so-called derivations. The simplest form of the rules, with obvious generalization to several sequents in the upper part of the rule, is

$$\frac{A_1, \ldots, A_r \leadsto B_1, \ldots, B_s}{A'_1, \ldots, A'_u \leadsto B'_1, \ldots, B'_v}$$

meaning: If one of $B_1, \ldots, B_s$ follows from all of $A_1, \ldots, A_r$ then one of $B'_1, \ldots, B'_v$ follows from all of $A'_1, \ldots, A'_u$.

The basis for a derivation are axioms (such as $P \leadsto P$), from which other sequents are arrived at by means of the rules.

The axioms can be of the *short* form $\Omega \leadsto \Omega$ or *extended* form $\Gamma, A \leadsto A, \Theta$.

The rules about negation are:

$$\frac{\Gamma \leadsto A, \Theta}{\Gamma, \neg A \leadsto \Theta} \qquad \frac{\Gamma, A \leadsto \Theta}{\Gamma \leadsto \neg A, \Theta}$$

The rules about disjunction are:

$$\frac{\Gamma, A \leadsto \Theta \quad \Delta, B \leadsto \Lambda}{\Gamma, \Delta, A \vee B \leadsto \Lambda, \Theta} \qquad \frac{\Gamma \leadsto A, B, \Theta}{\Gamma \leadsto A \vee B, \Theta}$$

The rules about conjunction are:

$$\frac{\Gamma, A, B \leadsto \Theta}{\Gamma, A \wedge B \leadsto \Theta} \qquad \frac{\Gamma \leadsto A, \Theta \quad \Delta \leadsto B, \Lambda}{\Gamma, \Delta \leadsto A \wedge B, \Lambda, \Theta}$$

The rules about implication are:

$$\frac{\Gamma \leadsto A, \Theta \quad \Delta, B \leadsto \Lambda}{\Gamma, \Delta, A \rightarrow B \leadsto \Lambda, \Theta} \qquad \frac{\Gamma, A \leadsto B, \Theta}{\Gamma \leadsto X \rightarrow Y, \Theta}$$

The rules about the universal quantifier are:

$$\frac{\Gamma, A[t] \leadsto \Theta}{\Gamma, \forall x A[x] \leadsto \Theta} \qquad \frac{\Gamma \leadsto A[u], \Theta}{\Gamma \leadsto \forall x A[x], \Theta}$$

where for right introduction, u should not occur in $\Gamma \leadsto \forall x A[x], \Theta$.

The rules about the existential quantifier are:

$$\frac{\Gamma, A[u] \leadsto \Theta}{\Gamma, \exists x A[x] \leadsto \Theta} \qquad \frac{\Gamma \leadsto A[t], \Theta}{\Gamma \leadsto \exists x A[x], \Theta}$$

where for left introduction, u should not occur in $\Gamma, \exists x A[x] \leadsto \Theta$.

Last, there is a rule for "reversing implication":

$$\frac{\Gamma, A \leadsto B, \Theta}{\Gamma, B \leadsto A, \Theta}$$

provided all axioms are in the short form.

The proviso in the rule is the reason behind the non-monotonicity of the system. An illustration can be given by deriving first $\forall x \forall y A(x) \vee B(y) \leadsto \forall z A(z) \wedge B(z)$ (bear in mind that such a sequent does *not* represent a tautological implication!). We start with the short form axiom

$$A(u), B(v) \leadsto A(u), B(v)$$

Then, we apply the rule for left introduction of conjunction

$$A(u) \wedge B(v) \leadsto A(u), B(v)$$

Next, we apply the rule for right introduction of disjunction

$$A(u) \wedge B(v) \leadsto A(u) \vee B(v)$$

As we have only short form axioms, the rule for "reversing implication" applies

$$A(u) \vee B(v) \leadsto A(u) \wedge B(v)$$

Using twice the rule for left introduction of the universal quantifier, we obtain

$$\forall x \forall y A(x) \vee B(y) \leadsto A(u) \wedge B(v)$$

By virtue of the rule for right introduction of the universal quantifier,

$$\forall x \forall y A(x) \vee B(y) \leadsto \forall z A(u) \wedge B(z)$$

and

$$\forall x \forall y A(x) \vee B(y) \leadsto \forall z \forall z A(z) \wedge B(z)$$

so that, by the usual condition on formulas of the language in first order logic,

$$\forall x \forall y A(x) \vee B(y) \leadsto \forall z A(z) \wedge B(z)$$

So, we have the formal case that $\forall z A(z) \wedge B(z)$ follows from $\forall x \forall y A(x) \vee B(y)$ but it is easy to see that we cannot obtain the same result from $C, \forall x \forall y A(x) \vee B(y)$. Indeed, C can only be introduced in the picture through an extended form axiom but this means that the step involving the rule for "reversing implication" is no longer allowed. In short, this shows that $\leadsto$ is non-monotonic because $\forall z A(z) \wedge B(z)$ follows from $\forall x \forall y A(x) \vee B(y)$ but not from $\forall x \forall y A(x) \vee B(y)$ supplemented with C.

7 Abductive reasoning with sequents

Let us give a simplistic example about abduction, together with the corresponding sequent derivation.

Consider a light bulb and a pocket battery. One wants to test whether the light bulb is ok. The background knowledge states that if the light bulb is ok then it lights whenever the battery is on. Should one observe that the light is on, one concludes that the battery is on and the light bulb is ok.

As for notation, we use the atoms *battery*, *light*, and *bulb* to denote that the battery is on, the light is on and the light bulb is ok, respectively. This means that the background theory T consists of the single formula *battery* $\wedge$ *bulb* $\rightarrow$ *light*, the manifestations M amount to the formula *light*, and the hypothesis H is the formula *battery* $\wedge$ *bulb*. The corresponding abductive reasoning can be formalized as follows. We start with the short form axioms

$$battery \wedge bulb \rightsquigarrow battery \wedge bulb$$

and

$$light \rightsquigarrow light$$

Then, we apply the rule for left introduction of implication

$$battery \wedge bulb \rightarrow light, battery \wedge bulb \rightsquigarrow light$$

All axioms are of the short form, the rule for "reversing implication" can apply

$$battery \wedge bulb \rightarrow light, light \rightsquigarrow battery \wedge bulb$$

That is, from the background knowledge, and from observing that the light is on, one infers that the battery is on and the light bulb is ok, as expected.

Of course, we can similarly handle more involved cases. Notably, we can deal with multiple causes for the same effect, we can deal with multiple observations, and so on. On the formal level, the same principle as above applies, the only difference is that the early stages in the derivations correspond to more complicated manipulations of the "classical rules" (the ones for first order logic) in the sequent system.

We have given an example of correct behaviour because it is the most interesting case – in other words, that's where abduction arises – in view of the above background knowledge (things would have been the other way around if the background knowledge would have been about incorrect behaviour). Still considering that all the background knowledge is about correct behaviour, what about the case where the observations provide evidence that something has gone wrong? Then, abduction is not needed, deduction is enough to yield the expected conclusion(s): In such a case, we only need to use the above sequent system as a classical deductive apparatus for first order logic – in the usual way.

In particular, returning to the above example, if it is observed that the light bulb does not light, then the sequent system yields the following derivation. We start with

$$battery \wedge bulb \rightsquigarrow battery \wedge bulb$$

and

$$light \rightsquigarrow light$$

Then, we apply the rule for left introduction of implication

$$battery \wedge bulb \rightarrow light, battery \wedge bulb \rightsquigarrow light$$

We now introduce negation to the left hand side

$$battery \wedge bulb \rightarrow light, battery \wedge bulb, \neg light \rightsquigarrow$$

and we introduce negation to the right hand side

$$battery \wedge bulb \rightarrow light, \neg light \rightsquigarrow \neg(battery \wedge bulb)$$

so that we obtain the desired conclusion.

Hence, we have established that from the background knowledge, it follows that: Observing that the light is not on implies that it's impossible for both the battery to be on and the light bulb to be ok.

Furthermore, from a subsequent (or coincidental) observation that the battery is indeed on (for instance, one may have verified it with one's tongue), then the sequent system admits a derivation stating that there is something wrong with the light bulb (i.e., the light bulb is not ok). The derivation, an easy one, is as follows. We start with the axioms of the extended form

$$battery, bulb \rightsquigarrow battery$$

and

$$battery, bulb \rightsquigarrow bulb$$

We apply the rule for left introduction of conjunction to obtain

$$battery, bulb \rightsquigarrow battery \wedge bulb$$

that we combine with the axiom

$$light \rightsquigarrow light$$

so that the rule for left introduction of implication gives

$$battery \wedge bulb \rightarrow light, battery, bulb \rightsquigarrow light$$

We continue with the rule for left introduction of negation

$$battery \wedge bulb \rightarrow light, battery, bulb, \neg light \rightsquigarrow$$

and then by the rule for right introduction of negation, we get

$$battery \wedge bulb \rightarrow light, battery, \neg light \rightsquigarrow \neg bulb$$

which is the expected result.

8   Inductive reasoning with sequents

Similarly to abduction, we give a simple illustration of applying the sequent system to induction.

We deal with the idea of mixing colours and we aim at learning about mixing two colours. We consider just a few colours: white, yellow, blue, red, green, purple. We use the self-explanatory symbols *white*, *yellow*, ... as first order constants and *mix* as a 3-place predicate where the last place corresponds to the colour resulting from mixing the colours indicated in the first two places.

First, let us assume that one observes that mixing blue and blue still yields blue. This is a positive example, there is no negative example. A first hypothesis that can be learned is that mixing a colour with itself does not change that colour.

In symbols, the examples M consist of the single formula $mix(blue, blue, blue)$ and H is then expected to be $\forall x\ mix(x, x, x)$. For the time being, the contents of the background theory are irrelevant and for simplicity, we assume that T is empty. Let us now give a derivation for the inductive reasoning of concluding H from M. We start with the short form axiom

$$mix(blue, blue, blue) \rightsquigarrow mix(blue, blue, blue)$$

We apply the rule for left introduction of the universal quantifier

$$\forall x\ mix(x, x, x) \rightsquigarrow mix(blue, blue, blue)$$

Then, we use the rule of "reversing implication"

$$mix(blue, blue, blue) \rightsquigarrow \forall x\ mix(x, x, x)$$

That is, we have a derivation of the sequent that relates the only example to the desired hypothesis.

Given at least one appropriate example, one can learn that the order of mixing does not change the result:

$$M' \rightsquigarrow \forall x \forall y \forall z\ (mix(x, y, z) \rightarrow mix(y, x, z))$$

or one can learn that white does not change the colour it is mixed with (we do not take into account colours fading):

$$M'' \rightsquigarrow \forall x\ mix(x, white, x)$$

If we are dissatisfied with learning that white does not change anything when mixed with another colour, we can introduce a negative example, say $mix(red, white, red)$ and maybe enrich the case with a related positive example $mix(red, white, pink)$. It must be observed here that adding that positive example does not dispense us to introduce the negative example. It would if the background theory contained the statement that mixing two colours always give a single colour and that pink is not the same colour as red.

A more interesting illustration is the case where one learns that mixing a colour with a composite one that contains it, still results in the composite one (once again, we do not take into account colours fading).

We use the examples where mixing red with purple yields purple whereas mixing red with blue yields purple. In symbols, M consists of $mix(red, blue, purple)$ and $mix(red, purple, purple)$, the background theory T is still assumed to be empty and H is expected to be $\forall x \forall y \forall z \, mix(x, y, z) \rightarrow mix(x, z, z)$.

We start with the short form axioms

$$mix(red, blue, purple) \rightsquigarrow mix(red, blue, purple)$$

and

$$mix(red, purple, purple) \rightsquigarrow mix(red, purple, purple)$$

so as to apply the rule for left introduction of implication, resulting in

$$mix(red, blue, purple), mix(red, blue, purple) \rightarrow mix(red, purple, purple)$$
$$\rightsquigarrow mix(red, purple, purple)$$

We apply three times the rule for left introduction of the universal quantifier

$$mix(red, blue, purple), \forall xyz \, mix(x, y, z) \rightarrow mix(x, z, z) \rightsquigarrow mix(red, purple, purple)$$

As all axioms are of the short form, we can use the rule for "reversing implication"

$$mix(red, blue, purple), mix(red, purple, purple) \rightsquigarrow \forall xyz \, mix(x, y, z) \rightarrow mix(x, z, z)$$

and we have a derivation of the expected sequent in which the desired hypothesis follows from the two given observations.

9 Conclusion

At first sight, even on a formal level, abduction and induction seem to be different reasoning patterns. However, they come under the same umbrella of non-monotonic inference. We have in fact shown that they are two syntactical specializations of the same scheme in a logic setting, a scheme linking hypotheses to observations (both to be found in abduction as well as induction). We have defined a sequent system that admits a formalization of non-monotonic inference from observations to hypotheses. We have applied it in the case of two small examples of diagnostic and learning.

Although we have provided firm grounds for an inferential account of abduction and induction in a logic setting, there still remains to deal with an important topic. Indeed, on top of the step of hypothesis formation that we have here taken care of, goes hypothesis confirmation which has been argued [5] to be the other main aspect when it comes to abductive and inductive reasoning.

10 Acknowledgements

This work was funded by the Commission of the European Union, ESPRIT III Basic Research Action Project DRUMS II.

References

1. F. Bergadano and D. Gunetti. An Interactive System to Learn Functional Logic Programs. In *Proc. 13th Int. Joint. Conf. on Artificial Intelligence* (IJCAI-93), pages 1044-1049, Chambéry, France, 1993. Morgan Kaufmann.

2. K. L. Clark. *Negation as Failure*. In *Logic and Databases*, (J. Minker, H. Gallaire, Eds.), pages 293-322, 1978. Plenum Press.

3. L. DeRaedt, N. Lavrac, and S. Dzeroski. Multiple Predicate Learning. In *Proc. 13th Int. Joint. Conf. on Artificial Intelligence* (IJCAI-93), pages 1037-1042, Chambéry, France, 1993. Morgan Kaufmann.

4. H.-D. Ebbinghaus, J. Flum and W. Thomas. *Mathematical Logic*. Springer Verlag, 1984.

5. P. A. Flach. A Model of Induction. In *Knowledge Representation and Reasoning under Uncertainty*, (M. Masuch, L. Polos, Eds.), Lecture Notes in Artificial Intelligence 808, pages 41-56, 1994. Springer Verlag.

6. J. H. Gallier. *Logic for Computer Science. Foundations of Automatic Theorem Proving*. Wiley, 1987.

7. G. Harman. *Change in View. Principles of Reasoning*. MIT Press, 1986.

8. B. Goedbhart. Abduction and Uncertainty in Compositional Reasoning. In *Proc. 11th European. Conf. on Artificial Intelligence* (ECAI-94), pages 70-74, Amsterdam, The Netherlands, 1994. Wiley.

9. R. Khardon and D. Roth. Reasoning with Models. In *Proc. 12th Nat. Conf. on Artificial Intelligence* (AAAI-94), pages 1148-1153, Seattle WA, 1994. Morgan Kaufmann.

10. S. Kraus, D. Lehmann and M. Magidor. Nonmonotonic Reasoning, Preferential Models and Cumulative Logics. *Artificial Intelligence*, 44: 167-207, 1994.

11. D. B. Leake. Focusing Construction and Selection of Abductive Hypotheses. In *Proc. 13th Int. Joint. Conf. on Artificial Intelligence* (IJCAI-93), pages 24-29, Chambéry, France, 1993. Morgan Kaufmann.

12. R. S. Michalski. A Theory and Methodology of Inductive Learning. *Artificial Intelligence*, 20:111-161, 1983.

13. S. Muggleton. Inductive Logic Programming. *New Generation Computing*, 8(4):295-318, 1991.

14. W. Narkiewicz. *Number Theory*. World Scientific, 1983.

15. G. Polya. *Mathematics and Plausible Reasoning*. Volume II: *Patterns of Plausible Inference*. Princeton University Press, 1954.

16. O. Raiman, J. de Kleer and V. Saraswat. Critical Reasoning. In *Proc. 13th Int. Joint. Conf. on Artificial Intelligence* (IJCAI-93), pages 18-23, Chambéry, France, 1993. Morgan Kaufmann.

INDEPENDENCE RELATIONSHIPS IN POSSIBILITY THEORY AND THEIR APPLICATION TO LEARNING BELIEF NETWORKS

L.M. de Campos
University of Granada, Granada, Spain

Abstract

This paper is devoted to the study of the concept of independence and conditional independence for possibility distributions. Different alternatives are considered, and their properties are analized with respect to a well-known set of axioms which try to capture the intuitive idea of independence. Moreover we carry out a formal study of the conditions that would lead to appropriate definitions of independence for possibilities.

1 Introduction

The concept of irrelevance or independence among events or variables has been identified as one of the most important to perform efficiently reasoning tasks in extensive and/or complex domains of knowledge. Dependency is a relation that establishes a potential change in our current belief about an event or a variable in a given domain, as a consequence of a specific change of our knowledge about another event or variable in this domain. Therefore, if a variable X is considered as independent of another variable Y, given a state of knowledge Z, then our belief about X will not change if we obtain additional information about Y. In other words, independence allows us to modularize the knowledge in such a way that we only need to consult the information that is relevant to the particular question which we are interested in, instead of having to explore a complete knowledge base. In this way, the reasoning systems taking into account the independence relationships gain in efficiency. This is the case, for example of belief or causal networks ([14, 15, 22]), which codify the independence relations by means of graphs.

In the framework of systems which handle uncertain knowledge, the concept of independence and conditional independence has been studied in-depth only for probability measures (see, for example [5, 13, 19]), although there are also some works for other theories of uncertain information ([2, 3, 18]), as well as works that consider the problem from an abstract point of view ([15, 16, 20, 23]).

The aim of this paper is to investigate several ways to define independence relationships in the framework of possibility theory [24, 8]. Some other works about the same topic have recently appeared [1, 10].

The paper is divided in 7 sections: in section 2 we introduce the properties that are usually required to an independence relation, which lead to a graphical representation. The concepts of marginalization and conditioning for possibility measures used through the paper are described in section 3. Section 4 studies different definitions of possibilistic independence, based on a view of possibility measures as particular cases of plausibility functions. In section 5 we do a brief study of the properties that a similarity relation between possibility distributions should verify, in order to obtain appropriate similarity-based definitions of independence. In section 6 other definitions of independence are considered, based on a different view of possibility measures, closer to fuzzy sets. Finally, section 7 contains the concluding remarks.

2   Axioms for Independence Relations

The study of the concept of conditional independence in probability theory (also in data base theory and graph theory) has resulted in the identification of several properties that could be reasonable to demand to any relation which tries to capture the intuitive notion of independence [15, 23]. If we denote by $I(X, Y | Z)$ the sentence 'X is independent of Y given Z', where X, Y y Z are (disjoint) sets of variables in a given domain, such properties are the following:

- A1 Trivial Independence: $I(X, \emptyset | Z)$

- A2 Symmetry: $I(X, Y | Z) \Rightarrow I(Y, X | Z)$

- A3 Decomposition: $I(X, Y \cup W | Z) \Rightarrow I(X, Y | Z)$

- A4 Weak Union: $I(X, Y \cup W | Z) \Rightarrow I(X, W | Z \cup Y)$

- A5 Contraction: $I(X, Y | Z)$ and $I(X, W | Z \cup Y) \Rightarrow I(X, Y \cup W | Z)$

- A6 Intersection: $I(X, Y | Z \cup W)$ and $I(X, W | Z \cup Y) \Rightarrow I(X, Y \cup W | Z)$

where X, Y, Z, W are arbitrary disjoint sets of variables.

The intuitive interpretation of these axioms is the following: Trivial independence says that our knowledge about X does not change if we do not obtain any additional information. Symmetry asserts that in any state of knowledge Z, if Y tell us nothing new about X, then X tell us nothing new about Y. Decomposition establishes that if two combined pieces of information Y and W are considered irrelevant to X, then each separate piece of information is also irrelevant. Weak union asserts that learning the irrelevant information Y cannot help the irrelevant information W become relevant to X. Contraction says that if two pieces of information, X and W, are irrelevant to each other after knowing an irrelevant information Y, then they were irrelevant before knowing Y too. Together, Weak union and Contraction mean that irrelevant information should not modify the character of being relevant or irrelevant of other propositions in the system. Finally, Intersection asserts that if two combined items of information Y and W are relevant to X, then at least one of them is also relevant to X when the other is added to our previous state of knowledge Z.

For example, in probability theory, and using the concepts of marginalization and conditioning which are usual for probabilities, independence is verified when the distribution of X conditioned to Y and Z and the distribution of X conditioned to Z are equal, that is

$$I(X, Y|Z) \Leftrightarrow P(x|yz) = P(x|z) \ \forall x \in X, \ \forall y, z \text{ such that } P(yz) > 0 \qquad (1)$$

This relation of probabilistic independence satisfies all the properties A1–A5, and it also verifies Intersection for probability distributions which are strictly positive.

Having an independence model M, that is, a set of conditional independence statements about the variables of a given domain, we try to find a graphical representation for it, or in other words, a correspondence between the variables in the model and the nodes in a graph G such that the topology of G reflects some properties of M. The kind of topological property of G which corresponds to the independence relations in M depends on the kind of graph: this property is separation if we choose a representation based on undirected graphs, and d-separation for directed acyclic graphs (see for example [15] for details). Depending on what properties are verified in each particular situation for a given independence model, it is possible to construct, using different methods, different types of graphs which display (through the separation or d-separation criteria) some or all the independences represented in the model [11, 15, 22].

After this brief review of the abstract aspects of independence and its relation with graphical structures, we are in position to study some specific formalisms. There are several theories to deal with numerical uncertainty in artificial intelligence. Some of them, at least in a formal sense, are hierarchycally ordered, going from the general to the specific. We have general fuzzy measures [21], lower and upper probabilities, Choquet capacities of order two [4], and belief and plausibility functions [6, 17], which include both possibility measures [24, 8] and probabilities. In the rest of the paper, we will focus in possibility theory, although some of the ideas presented here could be applied to more general theories.

3 Possibility Measures. Marginalization and Conditioning

Consider a finite referential $X = \{x_1, x_2, \ldots, x_n\}$ (or a variable taking its values on the set X). A possibility measure defined on X is a set function

$$\Pi : \mathcal{P}(X) \to [0, 1],$$

such that

1. $\Pi(X) = 1$,

2. $\Pi(A \cup B) = \max(\Pi(A), \Pi(B))$, $\forall A, B \subseteq X$.

Associated with the possibility measure Π, the possibility distribution

$$\pi : X \to [0, 1],$$

is defined by means of

$$\pi(x) = \Pi(\{x\}), \ \forall x \in X.$$

The information contained in the possibility distribution is sufficient to recover the possibility measure, since $\forall A \subseteq X$, $\Pi(A) = \max_{x \in A} \pi(x)$. So, we can restrict our study to deal only with possibility distributions.

Given a bidimensional possibility distribution π defined on the cartesian product $X \times Y$, the marginal possibility distribution on X, π_X (to simplify the notation, we will drop the subindex), is defined by means of

$$\pi_X(x) = \max_{y \in Y} \pi(x, y), \ \forall x \in X. \tag{2}$$

However, the concept of conditional possibility distribution is not so universal, there exist different alternatives to define it. We will mainly consider two alternatives:

1.- The possibility distribution on X conditioned to the event $[Y = y]$, $\pi_d(.|y)$, is defined as

$$\pi_d(x|y) = \frac{\pi(x, y)}{\pi(y)} = \frac{\pi(x, y)}{\max_{x' \in X} \pi(x', y)} \tag{3}$$

This is Dempster rule of conditioning, specialized to possibility measures, which are consonant plausibility functions ([17]).

2.- The possibility distribution on X conditioned to the event $[Y = y]$, $\pi_h(.|y)$, is defined as

$$\pi_h(x|y) = \begin{cases} \pi(x, y) & \text{if } \pi(x, y) < \pi(y) \\ 1 & \text{if } \pi(x, y) = \pi(y) \end{cases} \tag{4}$$

A slightly different version of this definition was considered in [12], as the solution of the equation $\pi(x, y) = \pi(x|y) \wedge \pi(y)$, $\forall x$. The definition above is the least specific solution of this equation [9].

Obviously these definitions of marginal and conditional possibility distributions can be inmediately extended to the n-dimensional case.

Observe that the difference between these two definitions of conditioning is the different kind of normalization used. In both cases we first focus on the values compatible with the available information $[Y = y]$, that is $\{(x, y'), \mid x \in X, y' = y\}$, and next we must normalize the values $\pi(x', y)$ to get a possibility distribution. In the first case the normalization is carried out by dividing by the greatest value $\max_{x' \in X} \pi(x', y)$ $(=\pi(y))$, whereas in the second case the normalization is achieved by increasing the greatest values of $\pi(x', y)$ up to 1.

4   Definitions of Independence Based on $\pi_d(.)$

Consider a finite set of variables $\mathcal{U}$, and a n-dimensional possibility distribution π defined on $\mathcal{U}$. Given three disjoint subsets of variables in $\mathcal{U}$, X,Y and Z, we will represent by means of $I(X, Y|Z)$ the sentence X is conditionally independent of Y given Z, for the model associated to the distribution π. We will denote by x, y, z the values that the respective variables can take on. The values of, for example, $Y \cup Z$, will be denoted by yz.

The most obvious way to define the conditional independence is to proceed in a way similar to the probabilistic case, that is to say, if the joint distribution of X, Y, Z factorizes. This is the idea considered in [18] for the more general framework of valuation-based systems. In our case this idea gives rise to:

$$I(X, Y|Z) \Leftrightarrow \pi_d(x|yz) = \pi_d(x|z), \tag{5}$$

that is to say, the knowledge about the value of the variable Y does not modify our belief about the values of X, when the value of variable Z is known; x, y, z are arbitrary values of the variables X, Y, Z, with the only restriction that the conditional distributions involved must be defined, that is to say, $\pi(yz) \neq 0$. This definition has been considered also in [23] for generalized conditional probabilities.

The above definition verifies the properties A1–A5, and it also satisfies A6 if the possibility distribution π is strictly positive.

One problem with this definition could be, in our opinion, that it is too strict, because it requires the equality of the distributions, and these distributions represent an imprecise knowledge. The problem becomes worse when the distributions must be estimated from data or human judgments.

The previous definition, eq.(5), represents that our information about X remains unchanged after conditioning to Y. A different idea could be to assert the independence when we do not gain additional information about the values of X after conditioning to Y (but we could lose some information). The idea of a possibility distribution being more or less informative than another one is adequately captured by the well-known definition of inclusion [7]: Given two possibility distributions π and π', π' is said to be included in (or gives less information than) π if and only if $\pi(x) \leq \pi'(x) \ \forall x$.

Using the relation of inclusion between possibility distributions, the previous idea can be expressed more formally by means of

$$I(X, Y|Z) \Leftrightarrow \pi_d(x|yz) \geq \pi_d(x|z). \tag{6}$$

Unfortunately this definition does not verify the property A4 (weak union), although it verifies A1–A3 and A5. We think that the problem is that the idea of independence as a non-gain of information has not been carried out until the end: if after conditioning we lose information, it seems more convenient to keep the initial information. This can be debatable, but it represents a sort of default rule: if in a very specific context we do not have much information, then to use the information available in a less specific context. In practical terms, this idea implies a change in the definition of conditioning:

$$\pi_{d_c}(x|y) = \begin{cases} \pi(x) & \text{if } \pi_d(x|y) \geq \pi(x) \ \forall x \\ \pi_d(x|y) & \text{if } \exists x' \text{ such that } \pi_d(x'|y) < \pi(x') \end{cases} \tag{7}$$

that is, if $\pi_d(x|y)$ is always greater than or equal to $\pi(x)$, that is, it gives less information than $\pi(x)$, then maintain $\pi(x)$; otherwise use the previous conditioning.

Using this conditioning, the new definition of independence is

$$I(X, Y|Z) \Leftrightarrow \pi_{d_c}(x|yz) = \pi_{d_c}(x|z). \tag{8}$$

This definition satisfies the properties A1 and A3–A6 (even the distributions which are not strictly positive verify Intersection). It does not satisfy Symmetry (we could obtain symmetry by defining a new relation $I'(.,.|.)$ by means of $I'(X, Y|Z) \Leftrightarrow I(X, Y|Z)$ and $I(Y, X|Z)$).

Another approach to define independence consists in considering that the possibility distributions $\pi_d(x|yz)$ and $\pi_d(x|z)$ are similar in some sense. So, if $\simeq$ is a relation in the set of possibility distributions, we could define the independence by means of

$$I(X, Y|Z) \Leftrightarrow \pi_d(x|yz) \simeq \pi_d(x|z). \tag{9}$$

In order to define the relation $\simeq$ we can consider different options. Let us see some of them:

If we think that a possibility distribution essentially establishes an ordering among the values that the variable we are considering can take on, and the numbers (the possibility degrees) only have a secondary importance, then we can say that two possibility distributions are similar if they establish the same order. More formally, we can define the relation $\simeq$ by means of

$$\pi \simeq \pi' \Leftrightarrow \forall x, x' [\pi(x) < \pi(x') \Leftrightarrow \pi'(x) < \pi'(x')]. \tag{10}$$

The independence obtained through this relation (we could call it 'isoordering') satisfies the properties A1 and A3–A5, and also A6 for strictly positive distributions; however it does not verify A2.

Another option is to speak about similarity between distributions when the possibility degrees of each distribution for each value are alike. More concretely, let m be any positive integer, and $\{\alpha_k\}_{k=0,\ldots,m}$ such that $\alpha_0 < \alpha_1 < \ldots < \alpha_{m-1} < \alpha_m$, with $\alpha_0 = 0$ and $\alpha_m = 1$. if we denote $I_k = [\alpha_{k-1}, \alpha_k)$, $k = 1, \ldots, m-1$, and $I_m = [\alpha_{m-1}, \alpha_m]$, then we define the relation $\simeq$ by means of

$$\pi \simeq \pi' \Leftrightarrow \forall x \; \exists k \in \{1, \ldots, m\} \text{ such that } \pi(x), \pi'(x) \in I_k. \tag{11}$$

This definition is equivalent to the following, which is established in terms of the α-cuts of the distribution:

$$\pi \simeq \pi' \Leftrightarrow C(\pi, \alpha_k) = C(\pi', \alpha_k) \; \forall k = 1, \ldots, m-1, \tag{12}$$

where $C(\pi, \alpha) = \{x \mid \pi(x) \geq \alpha\}$. This idea is also equivalent to discretize the interval $[0, 1]$, and to say that two distributions are similar if their respective discrete versions coincide. The independence defined in these terms satisfies the properties A1 and A3–A6 (although A6 only for strictly positive distributions), and again the symmetry fails.

Finally, a third option to define $\simeq$ is based on the following idea: consider a threshold α_0, and suppose that only from values greater than α_0 it is considered interesting to differentiate between the possibility degrees of two distributions; the values whose possibility degrees are below the threshold are not considered relevant. In terms of the α-cuts, this relation $\simeq$ can be expressed as follows:

$$\pi \simeq \pi' \Leftrightarrow C(\pi, \alpha) = C(\pi', \alpha), \; \forall \alpha \geq \alpha_0, \tag{13}$$

which is equivalent to

$$\pi \simeq \pi' \Leftrightarrow C(\pi, \alpha_0) = C(\pi', \alpha_0) \text{ and } \pi(x) = \pi'(x) \; \forall x \in C(\pi, \alpha_0). \tag{14}$$

Once again, the independence defined in terms of this relation $\simeq$ verifies the properties A1, A3–A5 (and A6 for distributions strictly positive), and it does not satisfy A2.

5 Sufficient Conditions about $\simeq$

In this section we continue the idea of using a similarity relation $\simeq$ between the conditional possibility distributions $\pi_d(x|yz)$ and $\pi_d(x|z)$ in order to define the predicate

$I(X, Y|Z)$, but now we try to study what kind of properties of $\simeq$ are sufficient in order to guarantee that the associated independence relation verifies some of the properties A1–A6.

First, it is obvious that A1 (Trivial Independence) will be fulfiled if $\simeq$ is reflexive. It is also evident that transitivity of $\simeq$ guarantees A5 (Contraction). If in addition $\simeq$ is symmetric, then it can be deduced that A3 (Decomposition) is verified if and only if A4 (Weak Union) is also verified. Therefore, it seems that the equivalence relations $\simeq$ are good candidates for defining independence.

A sufficient condition for the fulfilment of A3 is that $\simeq$ verify the following property:

Let $\{\pi_s\}$ a family of possibility distributions, and $\{\lambda_s\}$ a family of positive real numbers. Then

$$\frac{\pi_s}{\lambda_s} \simeq \pi \Rightarrow \frac{\max_s \pi_s}{\max_s \lambda_s} \simeq \pi. \tag{15}$$

Moreover, for the case of strictly positive distributions, the fulfilment of the property above also guarantees A6 (Intersection).

So, every possibilistic independence relation defined as in eq.(9), in terms of an equivalence relation $\simeq$ verifying (15), satisfies the properties A1, A3–A6 (although the last one only for strictly positive distributions). The only property excluded of this context is Symmetry; this is somewhat surprising, because this is one of the properties of independence seemingly more intuitive. Obviously, the three relations $\simeq$ defined in the previous section are equivalent relations and they verify (15).

6 Definitions of Independence Based on $\pi_h(.)$

In this section we are going to propose some definitions of independence based on $\pi_h(.)$, the other concept of conditioning considered in section 2. Using again the idea of independence as a non-gain of information after conditioning, we can define an independence relation as in eq.(6), by means of

$$I(X, Y|Z) \Leftrightarrow \pi_h(x|yz) \geq \pi_h(x|z). \tag{16}$$

In this case the definition above verifies the properties A1–A5 and does not verify A6 in general (so there is a difference between this definition using π_d and using π_h). Moreover this definition turns out to be equivalent to the following:

$$I(X, Y|Z) \Leftrightarrow \pi(xyz) = \pi(xz) \wedge \pi(yz). \tag{17}$$

If we consider the case of marginal independence ($Z = \emptyset$), from eq.(17) we obtain

$$I(X, Y|\emptyset) \Leftrightarrow \pi(xy) = \pi(x) \wedge \pi(z), \tag{18}$$

and we have the well-known concept of non-interactivity for possibility measures or fuzzy sets. So, our concept of independence is that we could call conditional non-interactivity.

Observe that for this kind of conditioning, π_h, the only necessary operation is comparison. So, we could easily consider possibility distributions taking values in sets different from the $[0,1]$ interval: it suffices to use a set $(\mathcal{L}, \preceq)$, where

$$\mathcal{L} = \{L_0, L_1, \ldots, L_n\}$$

and

$$L_0 \preceq L_1 \preceq \ldots \preceq L_n,$$

that is to say, a totally ordered set (for example, a set of linguistic labels), and then define possibility measures by means of

$$\Pi : \mathcal{P}(X) \to \mathcal{L}$$

verifying:

1. $\Pi(X) = L_n$.

2. $\Pi(A \cup B) = \vee_{\preceq}(\Pi(A), \Pi(B))$, $\forall A, B \subseteq X$,

where $\vee_{\preceq}$ is the maximum (supremum) operator associated to the order $\preceq$. We could speak about these generalized possibility measures as qualitative possibility measures. In these conditions we can define the conditioning and the independence in exactly the same way as before (eqs.(4) and (16)), obtaining the same properties.

A different way of extension (analogous to that one considered in section 5 for the other conditioning, eq.(9)) would consist in considering a relation $\approx$ in the set of possibility distributions, and then define the independence by means of

$$I(X, Y | Z) \Leftrightarrow \pi(xyz) \approx \pi(xz) \wedge \pi(yz). \tag{19}$$

In this case, a sufficient condition for this relation satisfying the axioms A1–A5, is that $\approx$ is an equivalence relation compatible with the marginalization and combination of possibility distributions (using the minimum as the combination operator)

On the other hand, and following the idea expressed in section 4, we could also define a new concept of conditioning, analogous to that of eq.(7), by means of

$$\pi_{h_c}(x|y) = \begin{cases} \pi(x) & \text{if } \pi_h(x|y) \geq \pi(x) \; \forall x \\ \pi_h(x|y) & \text{if } \exists x' \text{ such that } \pi_h(x'|y) < \pi(x') \end{cases} \tag{20}$$

Using this conditioning, the new definition of independence is

$$I(X, Y | Z) \Leftrightarrow \pi_{h_c}(x|yz) = \pi_{h_c}(x|z). \tag{21}$$

This definition is similar to that of eq.(8), but replacing π_{d_c} by π_{h_c}, and we get the same properties: A1–A6, except Symmetry. This relation is more restrictive than the conditional non-interactivity, since every independence statement with this definition implies the corresponding statement of conditional non-interactivity, but the converse is not true.

7 Concluding Remarks

The concept of independence, like for other formalisms of uncertainty management, is also important in possibility theory. In this paper several alternative definitions of possibilistic independence have been proposed. Most of them are based on the concept of conditioning, although using different formulations of this concept. We have also studied their properties with respect to a well-known set of axioms, which try to capture the intuitive idea of independence. It is obvious that a lot of things remain to be done; for example, to study the consequences of a non-symmetric definition of independence with respect to its graphical representation; to consider other points of view which are not necessarily based on the conditioning, or to study the problems of constructing independence graphs for possibilities and propagating information in these structures. The study of ways to extract possibility distributions from raw data, and merge this information with subjective judgments from experts is also interesting from a practical point of view.

Acknowledgements

This work has been supported by the DGICYT under Project PB92-0939 and by the European Economic Community under Project Esprit III b.r.a. 6156 (DRUMS II).

References

[1] Benferhat, S., Dubois, D. and Prade, H.: Expressing independence in a possibilistic framework and its application to default reasoning, in: Proceedings of the Eleventh ECAI Conference (1994), 150–154.

[2] Campos, L.M. de and Huete, J.F.: Independence concepts in upper and lower probabilities, in: Uncertainty in Intelligent Systems (Eds. B. Bouchon-Meunier, L. Valverde, R.R. Yager), North-Holland, Amsterdam 1993, 49–59.

[3] Campos, L.M. de and Huete, J.F.: Learning non probabilistic belief networks, in: Symbolic and Quantitative Approaches to Reasoning and Uncertainty, Lecture Notes in Computer Science 747 (Eds. M. Clarke, R. Kruse, S. Moral), Springer Verlag, Berlin 1993, 57–64.

[4] Choquet, G.: Theory of capacities, Ann. Inst. Fourier, 5 (1953), 131–292.

[5] Dawid, A.P.: Conditional independence in statistical theory, J.R. Statist. Soc. Ser. B, 41 (1979), 1–31.

[6] Dempster, A.P.: Upper and lower probabilities induced by a multivalued mapping, Ann. Math. Stat., 38 (1967), 325–339.

[7] Dubois, D. and Prade, H.: A set-theoretic view of belief functions, International Journal of General Systems, 12 (1986), 193–226.

[8] Dubois, D. and Prade, H.: Possibility Theory: An approach to computerized processing of uncertainty, Plenum Press, New York 1988.

[9] Dubois, D. and Prade, H.: Belief revision and updates in numerical formalisms– An overview, with new results for the possibilistic framework, in: Proceedings of the 13th IJCAI Conference, Morgan and Kaufmann 1993, 620–625.

[10] Farinas del Cerro, L. and Herzig, A.: Possibility theory and independence, in: Proceedings of the Fifth IPMU Conference 1994, 820–825.

[11] Spirtes, P., Glymour, C. and Scheines, R.: Causation, Prediction and Search. Lecture Notes in Statistics 81, Springer-Verlag 1993.

[12] Hisdal, E.: Conditional possibilities, independence and noninteraction. Fuzzy Sets and Systems, 1 (1978), 283–297.

[13] Lauritzen, S.L., Dawid, A.P., Larsen, B.N. and Leimer, H.G.: Independence properties of directed Markov fields, Networks, 20 (1990), 491–505.

[14] Pearl, J.: Fusion, propagation and structuring in belief networks, Artificial Intelligence, 29 (1986), 241–288.

[15] Pearl, J.: Probabilistic reasoning in intelligent systems: Networks of plausible inference, Morgan and Kaufmann, San Mateo 1988.

[16] Pearl, J., Geiger, D. and Verma, T.: Conditional independence and its representation, Kybernetika, 25 (1989), 33–44.

[17] Shafer, G.: A mathematical theory of evidence, Princeton University Press, Princeton 1976.

[18] Shenoy, P.P.: Conditional independence in uncertainty theories, in: Uncertainty in Artificial Intelligence, Proceedings of the Eighth Conference (Eds. D. Dubois, M.P. Wellman, B. D'Ambrosio, P. Smets), Morgan and Kaufmann, San Mateo 1992, 284–291.

[19] Spohn, W.: Stochastic independence, causal independence and shieldability, Journal of Philosophical Logic, 9 (1980), 73–99.

[20] Studený, M.: Attempts at axiomatic description of conditional independence, Kybernetika, 25 Supplement to n. 3 (1989), 72–79.

[21] Sugeno, M.: Theory of fuzzy integrals and its applications, Ph.D. Thesis, Tokio Institute of Technology, Tokio 1974.

[22] Verma, T. and Pearl, J.: Causal networks: Semantics and expressiveness, in: Uncertainty in Artificial Intelligence 4 (Eds. R.D. Shachter, T.S. Levitt, L.N. Kanal, J.F. Lemmer), North-Holland, Amsterdam (1990), 69–76.

[23] Wilson, N.: Generating graphoids from generalized conditional probability, in: Uncertainty in Artificial Intelligence, Proceedings of the Tenth Conference 1994.

[24] Zadeh, L.A.: Fuzzy sets as a basis for a theory of possibility, Fuzzy Sets and Systems, 1 (1978), 3–28.

USING ABDUCTION TO LEARN HORN THEORIES

D. Gunetti
University of Turin, Turin, Italy

Abstract

A method for learning Horn theories based on a systematic use of abduction is presented. Abduction is applied on the basis of the examples provided initially and of the hypothesis space to generate queries for missing positive and negative examples. The added examples are treated in the same way again and again until no more examples can be added. The process can be seen as a way of exploring the hypothesis space in order to build backward all the existing proof trees for the initial examples. The abductive completion procedure is used to solve the problems of extensional top-down learning methods. By means of abduction, a solution consistent with respect to the positive and negative examples given initially can always be found, if it exists in the hypothesis space.

1   Introduction

Learning can be seen as the process of going from an initial set of provided examples to a possible description of an unknown relation entailing the initial, and perhaps other, examples. A well known problem of this process is to provide the learning algorithm with an "adequate" set of examples of the target relation. Here "adequate" informally means that the training set should contain all those examples required to successfully complete the learning task, and no more. Obviously, this is a very hard condition to achieve, since usually it is not possible to know in advance exactly which examples are (and which are not) meaningful for learning a concept. As a consequence, the learning task may turn out to be too slow (if too many examples are given) and/or fails (if the examples are not meaningful).

This problem is particularly serious for an important class of learning methods: *Relational Learning Algorithms* based on an extensional interpretation of sub-predicates and recursion. In these methods, the learning procedure not only can fail or be too slow, it can also produce wrong results: the description of the target concept synthesized in the learning task may entail some of the negative examples given to the system. In this paper we show how abduction can be used to remedy the above problem. An abductive procedure is used to query the user for any example that may be missing, depending on the hypothesis space that has been defined and the examples given initially. A similar technique has been used before, for example in [9], to query the user for missing values allowing a single example to be covered. The novelty of the approach is that abduction is systematically applied over the whole hypothesis space. As a result of this, the learning systems turn out to be correct and sufficient, in the sense that a learned description will never entail any of the negative examples and such a description can always be found if it exists.

2   Relational Learning

Relational Learning Algorithms learn recursive relational descriptions from positive and negative examples, given as ground literals. Usually, a subset of the first-order predicate calculus, Horn clauses, is used for the concept description (see, e.g. [5, 25, 8, 3, 21]). Because of the use of Horn logic, the learned descriptions can be run as a normal logic program. For this reason, relational learning is also known as Inductive Logic Programming (ILP for short) [20].

The ILP problem can be described as follows: given a set of positive (E+) and negative (E-) examples of the target concept, a hypothesis space HS (a set of Horn clauses), and a background knowledge BK (consistent w.r.t. the negative examples but unable to explain the positive ones) find a subset P of HS such that:

1) $\forall$ e+ $\in$ E+ P $\cup$ BK $\vdash$ e+
2) $\forall$ e- $\in$ E- P $\cup$ BK $\nvdash$ e-

The first property is commonly known as *completeness*, and the second as *consistency*. If we suppose that an inductive method accepts as input a set of examples E and a description of a hypothesis space HS, the two properties defined below are also desirable:

Definition 1 *An induction procedure M is* correct *iff whenever M terminates successfully and outputs a program P belonging to HS, then P is complete and consistent w.r.t. E.*

Definition 2 *An induction procedure M is* sufficient *iff whenever a complete and consistent program w.r.t. E exists in HS, then M will output one such program.*

There are basicly two approaches to ILP: Bottom-up methods and Top-down methods. Bottom-up methods perform generalization steps starting from a description that is consistent but incomplete (usually, the empty set), until a solution complete, but still consistent, is found. These methods are based on the formal inversion of some deductive rule. There are methods based on *inverse unification* [22, 23] such as GOLEM [21], methods based on *inverse resolution* [19] such as CIGOL [19] and IRES [26], and methods based on *inverse implication* [12] such as LOPSTER [17] and CRUSTACEAN [1]. Bottom-up methods are theoretically well founded, but often inefficient. As a consequence, they are only able to learn simple concepts.

Top-down methods perform specialization steps starting from a description that is complete but inconsistent, until a solution consistent but still complete is found. To achieve efficiency, clauses are normally learned independently of each other, by adopting an extensional evaluation of the predicates and recursion, that will be described in the next section. Top-down methods have been emploied in real applications. Among the other, ILP techniques have ben used to learn rules for predicting the secondary structure of proteins from their amino acid sequence [15, 14]; to learn classification rules for early diagnosis of rheumatic diseases [18]; to automatically discover rules for relating the activity of drugs to their chemical structure and chemical properties of their subcomponents [13]; for learning diagnostic rules from qualitative models and discovering rules for qualitative reasoning [6]; for learning temporal diagnostic rules for physical systems [11] for the analysis of stresses in physical structures [10]; for learning qualitative models of dynamic systems [7]. Finally, ILP techniques have been successfully used as an alternative tool to the software engineering of Logic Programs [2].

3 Relational Learning Algorithms based on Extensionality

Learning definite clauses is difficult, especially in the case of multiple predicates, and systems tend to be slow. Many Top-down systems, such as Foil [24] and its derivatives, handle this problem evaluating clauses extensionally. In this way candidate clauses can be generated directly from the examples one at a time and independently of one another. As a consequence, the complexity of the learning task is proportional to the size of the hypothesis space, and not to its powerset. In the following, we give a brief description of the basic algorithm of the extensional approach.

Let P be the target concept and pos_examples(P) and neg_examples(P) the given positive and negative examples of P (in the following, α and γ represent generic conjunction of literals).

while pos_examples(P) is not empty do
    Learn_one _clause "P :- α";
    pos_examples(P) ← pos_examples(P) − pos(α)

Learn_one_clause:
α ← true;
while pos(α) is not empty do
    if neg(α) is empty then return(P :- α)
      else choose a predicate Q and its arguments Args;
          α ← $\alpha \wedge$ Q(Args)

where pos(α) and neg(α) are the sets of positive and negative examples of P covered by α. Every predicate Q can be defined by the user intensionally by means of logical rules or extensionally simply by giving some examples of its input-output behavior. In particular, clauses can be recursive and, in this case, Q = P, and its truth value can only be determined by the available examples.

We say that the clause P(X,Y) :- α(X,Y) *extensionally covers* an example P(a,b) iff:

- α = "Q(X,Y)". Then Q(X,Y) extensionally covers P(a,b) iff Q(a,b) is derivable from the definition of Q or is a given example of Q.

- α = "γ(X,T), Q(T,Y)" covers P(a,b) if there exists a value e such that the conjunction γ(a,e) is true and Q(e,b) is derivable from the definition of Q or is a given example of Q.

The choice of the literal Q(Args) to be added to the partial antecedent α of the clause being generated is guided by heuristic information. It might nevertheless be a wrong choice in some cases, in the sense that it causes the procedure "Generate one clause" to fail by exiting the while loop without returning any clause. This problem can be remedied by making the choice of Q(Args) a backtracking point.

Suppose, for instance, that learning system is given the following positive and negative tuples of the *ancestor* relation:

+ <r,g>,<b,g>,<b,d>,<d,g>,<b,r>,<r,d>,<r,p>;
− <d,d>,<g,p>,<d,p>.

where we also have an intensional definition for *parent*:

parent(X,Y) :- *mother*(X,Y)
parent(X,Y) :- *father*(X,Y)

where *mother* and *father* are (extensionally) defined by the following pairs of values:

mother		father	
X	Y	X	Y
d	g	f	g
r	d	s	d
r	p	s	p
i	r	b	r
t	s		
a	c	h	c

Finally, we know that the logic program for *ancestor* depends on *parent* and on itself (i.e. it may be recursive). As there are at most 3 variables to be used, these are the possible literals:

parent(X,Y), *parent*(Y,X), *parent*(X,W),
parent(W,X), *parent*(Y,W), *parent*(W,Y)
ancestor(X,W), *ancestor*(W,X), *ancestor*(Y,W), *ancestor*(W,Y)[1].

The learning algorithm starts to generate the first clause - the antecedent α is initially empty. We need to choose the first literal Q(Args) to be added to α. As we have left the heuristics unspecified, we will choose it so as to make the discussion short.
Let α=*parent*(W,Y); then all positive examples and the second and third negative examples are covered, so more literals need to be added.
Let α=*parent*(W,Y) $\wedge$ *parent*(W,X); in this case no positive examples are covered and the negative example <d,p> is covered. Clause generation fails and we backtrack to the last literal choice.
Let α=*parent*(W,Y) $\wedge$ *ancestor*(X,W); no negative example is covered, and the first 3 positive examples are extensionally covered. A clause is generated and the covered positive examples are deleted.
We proceed to the generation of another clause; α is empty again. If we choose the first literal as *parent*(X,Y), the remaining positive examples are covered and the final solution is obtained:

ancestor(X,Y) :- *parent*(W,Y), *ancestor*(X,W).
ancestor(X,Y) :- *parent*(X,Y).

The independence of the clauses is made possible by the extensional interpretation of recursion and sub-predicates: when a predicate Q occurs in a clause antecedent α, it is evaluated as true when the arguments match one of the positive examples. For instance, the clause

---

[1]*ancestor*(X,Y) and *ancestor*(Y,X) are not listed because they may produce looping recursions.

ancestor(X,Y) :- *parent*(W,Y), *ancestor*(X,W).

was found to extensionally cover the example <b,g> of *ancestor* because *parent*(d,g) is true, and <b,d> is also a positive example of *ancestor*. The previously generated logical definitions of Q are not used. The method is (partially) justified by the fact that if an example e is extensionally covered by a clause C $\in$ P, then P $\vdash$ e [3]. As a consequence of this, the learned description P is always *complete*, in the sense that derives every positive example provided initially.

However, extensionality forces us to include many examples, which would otherwise be unnecessary. In fact other desirable properties, similar to completeness, are not true, and two fundamental problems arise:

Problem 1: The learning task can fail. For a logic program P, it may happen that P $\vdash$ Q(q+), but none of its clauses extensionally cover q+. As a consequence an extensional system would be unable to generate P. Consider this program P:

ancestor(X,Y) :- *parent*(W,Y), *ancestor*(X,W).
ancestor(X,Y) :- *parent*(X,Y).

Let be given all the positive examples of *parent* and let <r,g> be the only given positive example of *ancestor*. This example follows from P (i.e., P $\vdash$ *ancestor*(r,g)) but it is not extensionally covered: the second clause does not cover it because *parent*(r,g) is false, and the first clause does not cover it extensionally because *parent*(d,g) is true, but <r,d> is not given as a positive tuple of *ancestor*. The learning task fails because a positive example is missing. Nonetheless, the provided example is perfectly representative of the concept we want to learn.

Problem 2: an inconsistent program can be learned. Even if no clause of a definition P extensionally covers a negative example q- of Q, it may still happen that P $\vdash$ Q(q-). Therefore an extensional system might generate a definition that covers negative examples that were given initially (i.e. extensional systems are, in general, not correct). Consider the following definition P:

ancestor(X,Y) :- *parent*(W,X), *ancestor*(W,Y)
ancestor(X,Y) :- *parent*(X,Y).

Then P $\vdash$ *ancestor*(g,p). Nevertheless, <g,p> is not extensionally covered by the first clause: *parent*(d,g) is true, but <d,p> is not a positive tuple of *ancestor*.
Since d is not an ancestor of p, it could not possibly be added as a positive tuple, and <g,p> would not be extensionally covered even if all positive tuples were given. The solution differs from the one of problem 1: in this case P will be ruled out by adding a *negative* example, namely <d,p>. In this way, the recursive clause can be found to

cover a negative example, and rejected.

Trying to solve the above problems by giving up the extensional interpretation of predicates while keeping the basic computational structure of extensional systems would produce insuperable problems. In fact, the truth value of the missing examples could be obtained by means of the partial program generated at a given moment. But if the inductive predicates occurring in a clause being generated are evaluated by means of the clauses that were learned previously, then the order we learn these clauses will becomes a major issue. In general, we must abandon our former assumption that clauses may be learned one at a time and independently. As a consequence, backtracking would be required. An alternative solution based on abduction is presented in the next section. The advantages of extensionality are preserved, i.e. previously generated clauses do not need to be reconsidered when learning new clauses.

4   Using abduction to learn

There is a naive solution to the problems of extensionality: providing the system with "many" examples of the target concept. However, this is likely to be ineffective. Most of the examples could be useless, but nevertheless they will have to be covered, slowing down the learning process and neutralizing the benefits of the extensional evaluation.

An alternative solution based on abduction is possible, in order to preserve the computational advantages of extensional approaches, while guaranteeing that a correct solution is found. The basic idea is starting with very few representative examples of the target concept, and using abduction to generate queries for the missing examples, on the basis of the given ones and of the hypothesis space. If we imagine all the positive and negative examples of the target concept available for answering the queries, abduction turns out to be used as a discriminating tool, in a fashion similar to the use made in abductive-EBL [8] to select the best explanation when more than one is available.

4.1   Existential queries through abduction

Consider the classical abductive inference rule:

$$\frac{\alpha \leftarrow \beta \qquad \alpha}{\beta}$$

that goes from observations to explanations, or from effects to causes, and let us adapt the rule to our framework, where:

- $\alpha \leftarrow \beta$ in the antecedent of the abductive rule represents a candidate clause,

- α represents a given example that can be matched with the head of the candidate clause,

- β in the consequent of the rule represents the (instantiated) body of the candidate clause after the match with the example.

Suppose that the instantiated body of the candidate clause cannot be evaluated to true on the basis of the known examples. Instead of concluding that the clause does not extensionally cover the example, we can generate a query for the missing examples to the user, or a search for those examples into a database of available examples of the target concept.

As an instance, let the example be P(a,b), and let

$$\alpha \leftarrow \beta = \text{P}(X,Y) :\text{-} \alpha(X,Y), Q(Y,Z).$$

Moreover, we know that $\alpha(a,b)$ is true (for example because every literal in α is extensionally evaluated to true). Then, that clause will extensionally cover P(a,b) only if there exists some value c such that Q(b,c) is a positive example (known to the system) of Q. Suppose that all such examples are missing. By adapting the abductive rule to our situation we get:

$$\frac{\text{P}(a,b) :\text{-} \alpha(a,b), Q(b,Z). \qquad \text{P}(a,b)}{Q(b,Z)}$$

and then the user can be queried for a value of Z such that Q(b,Z) is true. Alternatively, if a collection of examples of Q is available, a search for a positive example matching Q(b,Z) can be done. the new example of Q is added to the set of positive examples and P(a,b) can now be covered. This is a *controlled* form of abduction, in the sense that the result of abductive inference is not asserted as true, but only proposed as a possible truth, which is then queried to the user.

Existential abductive queries allow to cover uncovered positive examples, and partially solve the first problem of extensionality. They have been used before, for example in [28] and [9]. In the next section we show how apply these queries in a systematic way in order to completely solve the problems of extensional methods.

4.2 Exploring the hypothesis space through abduction

Let us apply the basic idea of generating abductive queries in a systematic way. That is, we want to build chains of abductive queries, or chains of controlled abductive inference steps. A backward reasoning from an initial example that we know to be true (and that must be extensionally covered), up to what must be true in order for

that initial example to be covered. The chain of abductive steps can be depicted as in
Figure 1.

$$\frac{\alpha \leftarrow \beta \quad \alpha}{\beta}$$

$$\Downarrow$$

$$\frac{\beta \leftarrow \gamma \quad \beta}{\gamma}$$

$$\Downarrow$$

$$\frac{\gamma \leftarrow \delta \quad \gamma}{\delta}$$

$$\Downarrow$$

$$\vdots$$

Figure 1: A chain of abductive steps

That is, a chain is produced because the queried examples are treated in the same
way as the starting one. The added examples are used to generate other abductive
queries again and again until no more examples can be added, and the backward
reasoning can stop. We show this on a very simple example.

Let be given the following hypothesis space HS containing only the correct clauses
for the *append* program:

{ c1 = app(X,Y,Z) :- head(H,H),tail(X,T),app(T,Y,W),cons(H,W,Z).
c2 = app(X,Y,Z) :- null(X),equal(Y,Z). }

Where all the predicates (except for *app*) are intensionally defined as usual. Let
the only given positive example be: $e_1 =$ "app([a,b],[c],[a,b,c])". A correct program
for *append* cannot be extensionally learned using only this example. However, let us
apply the abductive queries as described above. Example e_1 is matched against the
head of c1 producing the query "app([b],[c],W)?". That is, the user is queried for a
value of the variable W such that app([b],[c],W) is a positive example of *append*. Since
the answer is "W = [b,c]", the new example $e_2 =$ "app([b],[c],[b,c])" is added to the set
of known examples of *append* and treated in the same way. It is matched against c1

producing the new example e_3 = "app([],[c],[c])". Now, e_3 is handled in the same way, but it cannot be extensionally covered by c1, as the evaluation of "head([],H)" fails. So, it is matched with the head of clause c2 that is found to cover the example. Since no more examples are added and all the examples are extensionally covered by some clause in HS, the abductive reasoning can stop and so also the learning task. Clauses c1 and c2 have been learned.

In fact, the effect of building a chain of controlled abductive steps is that of producing a derivation tree in reverse order for the starting example e. That is, from e to the leaves of the tree, that will be clauses from the background knowledge, known (or queried) examples, and clauses from the hypothesis space (i.e. learned clauses).

If the same is done for each positive example given initially, and recursively for each queried positive example, then the first problem of extensionality is solved. If there exists a derivation tree for a positive example in the set {hypothesis space + background knowledge + known examples} such a tree is found, and together with it also the clauses necessary to the derivation. The found clauses belonging to the hypothesis space will be part of the description of the target concept.

However, the second problem of extensionality still remains. How can we avoid a (known) negative example to be derived by a learned description? In fact, the same strategy can be used, but in a complementary way. For each negative example, we test the existence of a derivation tree for it. If such a tree is found, clauses from the hypothesis space involved in the derivation will not be part of any description of the target concept. Let us show this with an example.

Consider the following (wrong) program P for *reverse*:

c1 = reverse(X,Y) :- head(X,H), tail(X,T1), head(Y,H), tail(Y,T2), reverse(T1,T2).
c2 = reverse(X,Y) :- null(X), null(Y).
c3 = reverse([X,Y,Z],[Z,Y,X]).

that can be learned by using this set of examples:

e_1 = reverse+([],[]), e_2 = reverse+([1],[1]), e_3 = reverse+([3,2,1],[1,2,3]).
e_4 = reverse-([3,2,1],[3,2,1]).

Then P ⊢ reverse([3,2,1],[3,2,1]), however, P is extensionally consistent with the provided examples. We can avoid learning P by means of a chain of abductive queries as follows: e_4 is matched against the head of c1 generating the query "reverse([2,1],[2,1])?" that is classified as a negative example (let call it e_5) by the user. Then, e_5 is matched against c1 and the clause is found to extensionally cover this example, since "reverse([1],[1])" is a known positive example. As a consequence, c1 extensionally covers a negative example (a queried one, indeed) and is rejected.

4.3   The abductive procedure

The ideas of the previous section are formalized in the abductive procedure below. Every clause C in the hypothesis space of the type:

$$C = P(X_1,..., X_m) :\text{-}\ \alpha,\ Q(W_1,..., W_n),\ \beta.$$

(where Q is a predicate defined extensionally - in particular, Q may be P, and α and β are conjunctions of literals), is processed as follows:

for every example $P(c_1,..., c_m)$ of P do
    match $P(c_1,..., c_m)$ against the head of C and
        propagate the instantiations to the body of C.
           (let these instantiations be represented by σ_1).
    evaluate the literals in $\alpha\sigma_1$ and
        propagate the instantiations to the rest of the body of C.
           (let each of the possible instantiations be represented by σ_2).
    ask the user for all the positive and
        negative examples of Q that match $Q(W_1,..., W_n)\sigma_1\sigma_2$[2];
        add these examples to the positive and negative examples of Q.

As we have seen, adding one example may cause the request of others. As a consequence, the procedure must be repeated for every clause, again and again, until no more examples are added.

Consider, for instance, the example about the *ancestor* relation given previously, and the following two clauses:

relative(X,Y) :- *ancestor*(X,Y).
ancestor(X,Y) :- *parent*(W,Y), *ancestor*(X,W).

where <b,g> is the only positive example of *relative*, <d,d> is the only positive example of *relative*, and <f,f> is the only negative example of *ancestor*; by using the first clause, the user is queried for *ancestor*(b,g), and this is added to the positive examples of *ancestor*. Using the second clause, *parent*(W,g) answers W_1=f and W_2=d; the user is then queried for the truth value of *ancestor*(b,d) and *ancestor*(b,f), that get added to the positive and negative examples, respectively. Since f has no parents, <b,f> does not cause the addition of more examples. By contrast, the second clause can be used again with X=b and Y=d, and, after some repetitions of this procedure, we obtain the completed set of examples for *ancestor*:

---

[2]Clearly, the procedure will terminate only if $Q(W_1,..., W_n)\sigma_1\sigma_2$ generates a finite number of queries, for each σ_2.

+ <b,g>,<b,d>,<b,r>
− <f,f>,<b,f>,<b,i>,<b,s>,<b,t>.

Not all possible examples have been added, only the ones that were useful for those two clauses, given the initial examples. If this is done for all the clauses that are allowed in the hypothesis space, then problems 1 and 2 do not longer hold:

Theorem: Suppose the examples given to an extensional learning system are completed with the above given abductive procedure. Suppose the learning system successfully exits its main loop and outputs a definition D for a concept P. If $D \vdash P(q)$ then $P(q)$ is extensionally covered by some clause in D.

Proof:
Suppose that (1) $D \vdash P(q)$ but (2) $P(q)$ is not extensionally covered.
Let "$P(X)$:- $\beta(X,Y),R(Y),\gamma$" be the clause resolved against $P(q)$. Suppose that $D \vdash \beta(q,r) \wedge \gamma$ and $D \vdash R(r)$, but no such r is a positive example of R. Such a literal $R(Y)$ must exists, because of assumptions (1) and (2). Since the tuples of R must have been completed with the given procedure, the user has been queried for $R(r)$, and this must have been inserted as a negative example. Therefore, it cannot be extensionally covered. We could now repeat the same argument for R. This would produce a non-terminating chain of resolution steps, and a finished proof of $P(q)$ would never be obtained, contradicting the hypothesis that $D \vdash P(q)$.

As a consequence of the theorem, if the examples given to an extensional learning system are completed with the above abductive procedure, and the learning system successfully exits its main loop and outputs a definition D, then D is consistent. If fact, if a negative example e^- - known to the system - is derived by D, then there must exist a clause in D extensionally covering e^-, because of the theorem. But then such a clause would have been rejected during the learning task.

It should be noted that this abductive completion is done as a preprocessing step, and it guarantees that a solution consistent with the examples is found if it exists. Moreover, it does not require reconsideration of previously generated clauses, as do systems (e.g. MIS [28]) that ask for new examples only when they are needed and during the learning process.

5   Discussion and Conclusion

One may wonder at which conditions the presented approach is practical. There are two main issues to consider. First, the abductive procedure terminates only if there is a finite number of answers to each question. This is the case if we are in a finite domain, as in the case of the *ancestor* relationship. In the case of infinite domains, the approach is still acceptable if we work with functional relations. By imposing adequate input/output constraints [3], it is possible to work only with positive examples. For each input tuple of the target relation, there is only one positive example matching the tuple, and infinite negative examples. the system can be instructed to work only with the positive examples, assuming the negative examples being all the examples with the same input as the positive ones and different output values. As a consequence, for each query, the user must provide exactly one positive example, and no more.

Second, the number of queries to the user must not only be finite, but also quite small (unless a database of examples is available, and the queries can be handled automatically). This depends mainly on the size of the hypothesis space, i.e. on the number of clauses that can be used to generate queries. In fact, it is becoming more and more clear as learning complex Horn description is quite difficult, and a lot of knowledge must be used. As a consequence, the user must not be a passive agent, just providing the examples of the target relation. She or he must also be able to design a restricted hypothesis space on the basis of the sought relation [4, 16]. Also the number of examples provided initially may influence the number of queries. Experiments based on the presented approach have shown as it is possible to learn complex relations by using very few representative examples of the target concept, and letting the abductive procedure asking for the missing ones. For example, it was possible to learn a program for inserting a key into a balanced tree (rebalancing the tree if necessary) by using an extensional approach and just one well chosen initial positive example. With a hypothesis space of 2^{24} clauses the user was queried for 15 more examples, and the program could be learned in 1831 seconds on a Sun Sparcstation 5. The program was composed of 9 clauses, eight of them recursive. In general, it can be shown that an initial positive example is sufficient to learn all the clauses necessary to derive it, if the abductive completion is applied.

In this work we have faced one of the most frequent kind of imperfect data met in Relational Learning: training examples of the target concept which are too sparse to allow a reliable correlation among them to be found. An abductive strategy has been presented able to explore the hypothesis space on the basis of the examples provided initially, in order to find a correct solution, whenever it exists. As a consequence of the presented approach, the user is not compelled to provide all the needed examples (which are, in general, unknown) to learn a definition, at a time. She or he can forget some examples, and the abductive procedure will ask for them. Only the examples really needed are queried, so the system will not waste time trying to cover useless examples. In many extensional systems the user must provide all the examples at one time, and usually a superset of the examples needed is given, resulting in a lot of time wasted in covering useless examples.

Acknowledgement: The author thanks esprit project DRUMS II which is currently supporting part of this research.

References

[1] D. W. Aha, S. Lapointe, C. X. Ling, and S. Matwin. Learning Recursive Relations with Randomly Selected Small Training Sets. In William W. Cohen, editor, *Proc. of the Int. Conf. on Machine Learning*, pages 12–18, New Brunswick, New Jersey, 1994. Morgan Kaufmann.

[2] F. Bergadano and D. Gunetti. Inductive Synthesis of Logic Programs and Inductive Logic Programming. In *Proc. Int. Workshop on Logic Program Synthesis and Transformation (LOPSTR-93)*, Leuven, Belgium, 1993. Springer Verlag, LNCS, in press.

[3] F. Bergadano and D. Gunetti. An interactive system to learn functional logic programs. In *Proc. 13th Int. Joint. Conf. on Artificial Intelligence*, pages 1044–1049, Chambery, France, 1993. Morgan Kaufmann.

[4] F. Bergadano and D. Gunetti. *Inductive Logic Programming: from Machine Learning to Software Engineering*. MIT Press, Cambridge, MA, 1995.

[5] L. Birnbaum and G. Collins, editors. *Proc. of the 8th Int. conference on Machine Learning, part VI: Learning Relations*. Morgan Kaufmann, 1991.

[6] I. Bratko. Qualitative modelling: learning and control. In *Proc. IJCAI*, Prague, Czechoslovakia, 1991.

[7] I. Bratko, S. Muggleton, and A. Varsek. Learning qualitative models of dynamic systems. In S. Muggleton, editor, *Inductive Logic Programming*, pages 437–452. Academic Press, London, 1992.

[8] W. Cohen. Abductive Explanation-Based Learning: a Solution to the Multiple Inconsistent Explanation Problem. *Machine Learning*, 8:167–219, 1992.

[9] L. DeRaedt and M. Bruynooghe. CLINT: a Multistrategy Interactive Concept-Learner and Theory Revision System. In R. S. Michalski and G. Tecuci, editors, *Proc. Workshop on Multistrategy Learning*, pages 175–190, Harpers Ferry, VA, 1991.

[10] S. Dzeroski and B. Dolsak. Comparison of ilp systems on the problem of finite element mesh design. In *Proc. 6th ISSEK Workshop*, Jozef Stefan Inst., Ljubljana, Slovenia, 1991.

[11] C. Feng. Inducing temporal fault diagnostic rules from a qualitative model. In S. Muggleton, editor, *Inductive Logic Programming*, pages 471–493. Academic Press, London, 1992.

[12] P. Idestam-Almquist. *Generalization of Clauses*. Ph.D. Thesis, Stockholm University, Sweden, 1993.

[13] R. King, S. Muggleton, R. Lewis, and M. Sternberg. Drug design by machine learning: The use of inductive logic programming to model the structure-activity relationships of trimethoprim analogues binding to dihydrofolate reductase. *Proc. National Academy of Sciences, USA*, 89:11322–11326, 1992.

[14] R. King, S. Muggleton, and M. Sternberg. Protein secondary structure prediction using logic. In *Proc. 2nd Int. Workshop on Inductive Logic Programming*, Tokio, Japan, 1992.

[15] R. King and M. Sternberg. Machine Learning approach for the prediction of protein secondary structure. *Journal of Molecolary Biology*, 216:441–457, 1990.

[16] W. Van Laer, L. Dehaspe, and L. DeRaedt. Applications of a logical discovery engine. Leuven, Belgium, 1994. Technical Report, Dept. of CS, Univ. of Leuven.

[17] S. Lapointe and S. Matwin. Sub-unification: a tool for efficient induction of recursive programs. In D. Sleeman and P. Edwards, editors, *Proc. of the Int. Conf. on Machine Learning*, pages 273–281, Aberdeen, Scotland, 1992. Morgan Kaufmann.

[18] N. Lavrac, S. Dzeroski, V. Pirnat, and V. Krizman. Learning rules for early diagnoses of rheumatic diseases. In *Proc. 3rd Scandinavian Conf. on Artificial Intelligence*, pages 138–149. IOS Press, Amsterdam, 1991.

[19] S. Muggleton. Machine Invention of First Order Predicates by Inverting Resolution. In *Proc. of the Fifth Int. Conf. on Machine Learning*, pages 339–352, Ann Arbor, MI, 1988.

[20] S. Muggleton, editor. *Inductive Logic Programming*. Academic Press, 1991.

[21] S. Muggleton and C. Feng. Efficient Induction of Logic Programs. In *Proc. of the first conf. on Algorithmic Learning Theory*, Tokyo, Japan, 1990. OHMSHA.

[22] G. Plotkin. A note on Inductive Generalization. In B. Meltzer and D. Michie, editors, *Machine Intelligence 5*, pages 153–163, 1970.

[23] G. Plotkin. A further note on Inductive Generalization. In B. Meltzer and D. Michie, editors, *Machine Intelligence 6*, pages 101–124, 1971.

[24] R. Quinlan. Learning Logical Definitions from Relations. *Machine Learning*, 5:239–266, 1990.

[25] C. Rouveirol, editor. *Proc. of the ECAI Workshop on Logical Approaches to Learning*. ECCAI, Vienna, Austria, 1992.

[26] C. Rouveirol and J. F. Puget. Beyond Inversion of Resolution. In Porter and Mooney, editors, *Proc. of the Int. Conf. on Machine Learning*, pages 122–130, 1990.

[27] E. Y. Shapiro. An Algorithm that Infers Theories from Facts. In *Proc. of the 7th Int. Joint Conf. on Artificial Intelligence*, pages 446–451, Vancouver, Canada, 1981.

[28] E. Y. Shapiro. *Algorithmic Program Debugging*. MIT Press, 1983.

ON LOGICS OF APPROXIMATE REASONING II

P. Hájek
Academy of Sciences, Prague, Czech Republic

ABSTRACT.

A logical analysis of reasoning under uncertainty and vagueness, surveyed in Part I of this paper, is continued and updated. The paper has a survey character.

0. INTRODUCTION.

This is a free continuation of my paper [1] (reprinted with the permission of Springer-Verlag as [2]); it will be helpful for the reader to have a copy of [1] or [2], thereafter called "Part I", at his/her disposal.

In Part I we paid much attention to the distinction of logics of vagueness (fuzzy logics) on the one side and logics of beliefs (uncertainty) on the other. Formally the former leads to many-valued logics and the latter to various modal logics in the style of Kripke models. The most apparent difference consists in the fact that fuzzy logics are truth-functional (extensional) whereas logics of belief are not. In the present paper we shall concentrate to logics of beliefs (and then to logics of both beliefs and vagueness); concerning fuzzy logics, the reader's attention is called to my paper [3] containing a substantially simplified formal systems of Pavelka-style fuzzy logic with the corresponding completeness theorem and a recursion-theoretic analysis as well as to my survery paper [4] stressing the symbolic (formal) side of fuzzy logic.

Thus the paper may be understood as an updating commentary to Sections 5-8 of Part I and as a partial substitute for an expected more extensive paper, which I have been unable to produce due to lack of time.

1. MODAL LOGICS.

In Part I we gave three examples of modal logics: (S5), called often the *logic of knowledge*; (L) (or GL), called *provability logic*, and a *tense logic* FLOT with finite (strictly) linearly ordered time (with only one necessity – always in future). This logic was modified to a system called FLPOT-tense logic with finite (strictly) linearly preordered time. It is convenient to extend this choice of examples with the following ones.

Example 4. Logic of belief (KD45)

propositional axioms

$\Box(A \to B) \to (\Box \to \Box B)$

$\Box A \equiv \Box \Box A$

$\Diamond A \equiv \Box \Diamond A$

rules: *modus ponens*

(from A and $A \to B$ infer B)

necessitation

(from A infer $\Box A$)

Note that (S5) is just the extension of (KD45) by the axiom $\Box A \to A$. We have the following:

Completeness theorem: S5 is sound and complete w.r.t. Kripke models $K = \langle W, \Vdash, R \rangle$ where $R = W \times W$

KD45 is sound and complete w.r.t. Kripke models $K = \langle W, \Vdash, R \rangle$ such that for some non-empty $W_0 \subseteq W$, $R = W \times W_0$.

Finite model property: If Λ is satisfiable (true in a world of a Kripke model) then A is satisfiable in a finite Kripke model (W finite).

Note that both S5 and KD45 have the finite model property.

See [5] for information on KD45. Note that the W_0 above is the set of "believable" worlds; $\Box A$ is true if A is true in all believable worlds.

Example 5. Reflexive tense logic (RTL) with non-strictly linearly preordered time. There are two necessities: $\hat{G}$ – always from now on $\hat{H}$ – always until now, and two possibilities: $\hat{F} = \neg \hat{G} \neg$, $\hat{P} = \neg \hat{H} \neg$.

Axioms: $\hat{G}(A \to B) \to (\hat{G}A \to \hat{G}B)$

$\hat{G}A \to \hat{G}\hat{G}A$

$\hat{G}A \to A$

the same for $\hat{H}$

$\hat{P}\hat{G}A \to A, \hat{F}\hat{H}A \to A$

$\hat{F}A \to \hat{G}(\hat{P}A \vee \hat{F}A)$ (not branching)

$\hat{P}A \rightarrow \hat{H}(\hat{P}A \vee \hat{F}A)$ (not branching)

Kripke models have the form $K = \langle W, \Vdash, R \rangle$ where R is a reflexive linear preorder (transitive and connected: $(wRv \vee vRw)$).

$w \Vdash \hat{G}A$ iff for all w', wRw' implies $w' \Vdash A$

$w \Vdash \hat{H}A$ iff for all w', $w'Rw$ implies $w' \Vdash A$

The logic RTL is complete and has finite model property.

2. UNCERTAINTY AND MODAL LOGIC.

We recall the basic setting: In the sequel W stands for a non-empty finite set of *possible worlds* and $\Vdash$ is a truth evaluation assigning to each proppositional atom P and each possible world w a truth value 1 or 0 - the truth-value of p in the world w.

A *probabilistic Kripke model* has the form $K = \langle W, \Vdash, P \rangle$ where $P : W \rightarrow [0,1]$(unit interval) and $\Sigma_{w \in W} P(w) = 1$. A *possibilistic Kripke model* has the form $K = \langle W, \Vdash, \pi \rangle$ where $\pi : W \rightarrow [0,1]$ and $max_{w \in W} \pi(w) = 1$. A *DS-Kripke model* (DS for Dempster and Shafer) has the form $K = \langle W, \Vdash, m \rangle$ where $m(A) \in [0,1]$ for each $A \subseteq W$ and $\Sigma_{A \subseteq W} m(A) = 1$.

A possibilistic model is *positive* if $\pi(w) > 0$ for each w. Note that a possibilistic model is a generalization of at model of KD45: instead of having a crisp subset $W_0 \subseteq W$ of believable worlds we have a fuzzy subset π of W; $\pi(w)$ is the degree of possibility of w.

One defines, for $A \subseteq W$, $P(A) = \Sigma_{w \in A} P(w)$, $\Pi(A) = max_{w \in A} \pi(w)$, $bel(A) = \Sigma_{\emptyset \neq B \subseteq A} m(B)$ and defines the probability (possibility, belief) of a formula to be the corresponding measure of the set of all possible worlds in which the formula is true.

There are reasonable generalizations for infinite Kripke models (W being infinite); a probability is then defined as a function on an algebra (or σ - algebra) of subsets of W satisfying the usual conditions (for a finitely or countably additive measure), the generalization for possibility is obvious (only replace max by sup). An elegant definition of belief functions on an infinite frame is found in [6].

Example

Here we have a probabilistic, possibilistic and DS-Kripke model, all three models having the same set W and truth assignment $\Vdash$.

	p q	P	Π
1	1 1	0.3	0.3
2	1 0	0.2	0.7
3	0 1	0.0	1.0
4	0 0	0.4	1.0
5	1 1	0.1	1.0

$m(\{1,3,4,5\}) = 0.7$

$m(\{1,2,3,4,5\}) = 0.3$

$P(\{1,2\}) = 0.5$

$\Pi(\{1,2\}) = 0.7, N(\{1,2\}) = 0$

$bel(\{1,2\}) = 0, pl(\{1,2\}) = 1$
$bel(\{1,3,4,5\}) = 0.7$, etc.

In Part I we refered to the work of Ruspini on similarities and to the work of Klir et al. on belief as probability of necessity. We give a complete information on the latter.

Consider Kripke models (finite)

$$K = \langle W, \Vdash, P, R \rangle$$

where P is a probability on W and $R \subseteq W \times W$. Define

$$bel(A) = P(\{w | w \Vdash \Box A\})$$

(probability of necessity / provability).

Theorem: (1) bel is a belief function; if R is reflexive and P positive then bel is regular ($bel(W) = 1$).
(2) For each $K = \langle W, \Vdash, bel \rangle$ where bel is a regular belief function there is a
$K' = \langle W', val', P, R \rangle$ where R is an equivalence (thus $\langle W', val', R \rangle$ is a model of S5) and for each formula A,

$$bel_K(A) = bel_{K'}(A).$$

(3) For each $K = \langle W, \Vdash, bel \rangle$ where bel is an arbitrary belief function there is a
$K' = \langle W', \Vdash', P, R \rangle$ where R is a strict partial order (thus $\langle W', \Vdash', R \rangle$ is a model of provability logic) and, for each formula A,

$$bel_K(A) = bel_{K'}(A)$$

(belief as probability of provability).

For (1) and (2) see [7], for (3) see [8]. (3) has some appeal since it defines belief as probability of provability in the sense of the modal provability logic. Belief as probability (in a non-modal, general sense) was steressed by several authors, e.g. [9], [10].

In Part I we referred on a *comparative possibilistic logic QPL^+* whose models are possibilistic Kripke models and which has a binary modality $\lhd$ of comparison of possibilities: $K \Vdash A \lhd B$ iff $\Pi_K(A) < \Pi_K(B)$ (where Π_K is the possibility given by K). We referred to [11] for a complete axiomatization and to a simplification of Herzig. It has turned out that Herzig's system does not do the job but a small modification of it does.

Axioms:
$(A \lhd B) \& (B \lhd C) \to (A \lhd C)$ (transitivity)
$(A \lhd B) \lor (B \lhd A)$ (linearity)
$\neg(True \lhd False)$ (non-triviality)
$(A \lhd True)$
$(A \lhd B) \to ((A \lor C) \lhd (B \lor C))$ (monotonicity)

def.: $\Box A$ is $\neg A \lhd False$

$(A \lhd B) \rightarrow \Box(A \lhd B)$

$\neg(A \lhd B) \rightarrow \Box\neg(A \lhd B)$

Rules: Modus ponens and necessitation: from $A \rightarrow B$ infer $A \lhd B$.

QPL_{pos} is QPL plus $\Box A \rightarrow A$.

Theorem: (1) QPL is complete for all finite possibilistic Kripke models $K = \langle W, \Vdash, \Pi \rangle$, $\Pi : W \rightarrow [0,1]$, $\max_w \Pi(w) = 1$. QPL_{pos} is complete for all **positive** such models ($\Pi(w) > 0$ for each w).

(2) The translation of KD45-formulas into QPL-formulas given by

$$(\Box A)^* \text{ is } \neg A^* \lhd False$$

is a faithful embedding of KD45 into QPL and of S5 in QPL_{pos}.

(3) The translation of QPL-formulas into the reflexive tense logic RTL given by

$$(A \lhd B)^* \text{ is } \Box(A \rightarrow \hat{F}B)$$

is a faithful embedding of QPL_{pos} into RTL

Now let as turn to comparative belief logic (in the sense of Dempster-Shafer theory). The following definition (of a logic QBL) and theorem are from [12].

Axioms:

$(A \lhd B)\&(B \lhd C) \rightarrow (A \lhd C)$ transitivity

$(A \lhd B) \vee (B \lhd A)$ linearity

$\neg(True \lhd False)$ non-triviality

definition: $\Box A$ is $True \lhd A$

$\Box(A \rightarrow B) \rightarrow (A \lhd B)$

$\Box(A \rightarrow B)\&\neg\Diamond(A\&C)\&((B \vee C) \lhd (A \vee C)) \rightarrow B \lhd A$

$(A \lhd B) \rightarrow \Box(A \lhd B)$

$\neg(A \lhd B) \rightarrow \Box(\neg A \lhd B)$

Theorem: QBL is complete for all finite belief models $K = \langle W, \Vdash, bel \rangle$.

The embedding $*$ of KD45-formulas into QBL-formulas given by

$$\Box A \text{ is } (True \lhd A)$$

is a faithful embedding of KD45 into QBL.

Remark. Note that the comparative possibilistic logic is not just a strengthening of the comparative belief logic (necessity is defined in a different way); this is because possibility is a particular plausibility (dual notion of belief, $pl(A) = 1 - bel(W - A)$), not a particular belief, see Part I.

3. VAGUENESS AND UNCERTAINTY; MANY-VALUED MODAL LOGICS.

Here we briefly refer on two approaches to the problem of building a comparative fuzzy possibilistic logic, i.e. a many-valued modal logic with a notion of numeric possibility of a formula. One is based on a fuzzy analogy of (S5) and (KD45) - see [13], another one on a fuzzy tense logic - see [14].

The whole discussion takes a finitely-valued Lukasiewicz's logic with the set

$$Values = \{0, \frac{1}{n-1}, \frac{2}{n-1}, \ldots, 1\}$$

as the set of truth values and with usual connectives of Lukasiewicz's logic enriched by *coefficients* (i) for each $i \in Val$:

$$\| (i)A \| = \begin{cases} 1 & \text{if } \| A \| = i \\ 0 & \text{otherwise} \end{cases}$$

Note that the coefficients are definable from other connectives, but in a cumbersome way.

Boolean formulas result from formulas of the form $(i)A$ using connectives and modalities (if any). For boolean formulas we have classical logic.

For details see [13], [14] and references thereof.

We first present our variant of fuzzy S5 and KD45.

The theory MVS5 (many-valued S5) has the following axioms (cf.[13]):
axioms of many-valued propositional logic,
axioms of S5, i.e.
$\Box(A \to B) \to (\Box A \to \Box B)$,
$\Box A \equiv \Box\Box A$,
$\Diamond A \equiv \Box\Diamond A, \Box A \to A$, and
$(\geq i)\Box A \equiv \Box(\geq i)A$
(where $(\geq i)B$ is $\bigvee_{j \geq i}(j)B$)

Models have the form $K = \langle W, val, W \times W \rangle$ where $val(p, w) \in Values$ for each atom p and posible world w, values of formulas are defined in the obvious way, e.g.
$\| p \|_w = val(p, w), \| \neg A \|_w = 1 - \| A \|_w,$

. . .

$\| \Box A \|_w = \min\{\| A \|_{w'} \,|\, w' \in W\}$

A is a 1-*tautology* iff $\| A \|_w = 1$ for all w in all K.

MVS5 is *complete*, i.e. a formula is a 1-tautology iff it is provable from the axioms using some evident deduction rules (modus ponens, necessitation and some auxiliary ones).

MVKD45 is formulated in a similar manner but the axiom system is slightly cumbersome: we have propositional axioms, axioms of KD45 and five additional schemes, e.g.

$$(j)\Box A \equiv \Box(j)\Box A$$

Models have the forms $K = \langle W, val, \pi \rangle$ where $val : Prop\_var \times W \to Values,$ $\pi : W \to Values, \max_w \pi(w) = 1$

We define
$$\| \Diamond A \|_w = \max_w(\| A \|_w \wedge \pi(w))$$
$$\| \Box A \|_w = \| \neg \Diamond \neg A \|_w$$
This logic is complete in the usual sense.

Now we can present our first definition of the possibility of a formula A in this fuzzy logic:
$$\Pi(A) = \| \Diamond A \| = \max_w(\| A \|_w \wedge \pi(w))$$

This leads to a comparative fuzzy possibilistic logic CFPL with a binary modality of comparison of possibilities defined as follows.

$$\|A \lhd B\|_w = \|\Diamond A \rightarrow \Diamond B\|_w$$

Fact: (1) $\|A \lhd B\| = 1$ iff $\Pi(A) \leq \Pi(B)$
(2) The mapping $*$ replacing $A \lhd B$ by $\Diamond A \rightarrow \Diamond B$ and commuting with connectives is a faithful embedding of this CFPL into MVKD45.

Remark: CFPL can be also related to MVS5 using an extra propositional variable.

Our second approach is related to a fuzzy tense logic MTL with two modalities $\hat{G}$ (always from now on), $\hat{H}$ (always until now) and their duals $\hat{F}, \hat{P}$, and the following axioms.

Axioms: fuzzy propositional,
$\hat{G}(A \rightarrow B) \rightarrow (\hat{G}A \rightarrow \hat{G}B)$
$\hat{P}\hat{G}A \rightarrow A$
$\hat{G}A \rightarrow A$
$\hat{G}A \rightarrow \hat{G}\hat{G}A$
$\hat{F}A \rightarrow \hat{G}(\hat{P}A \vee \hat{F}A)$
$\hat{G}(\geq i)A \equiv (\geq i)\hat{G}A$
$\hat{G}(\geq i)A \equiv (\geq i)\hat{F}A$
and duals: replace $\hat{G}, \hat{F}, \hat{P}$ by $\hat{H}, \hat{P}, \hat{F}$ respectively

Kripke models have the form $K = \langle W, val, \leq \rangle$ where $\leq$ is a reflexive linear preorder;

$$\|\hat{G}A\|_w = \min_{w' \geq w} \|A\|_w$$

$$\|\hat{H}A\|_w = \min_{w' \leq w} \|A\|_w$$

We have completeness and finite model property.
Now investigate fuzzy possibilistic models $K = \langle W, val, \pi \rangle$ where

$$\pi : W \rightarrow [0,1], \quad \sup_w \pi(w) = 1$$

The possibility of a formula A is now defined as follows (second choice): put

$$\sigma_A(i) = \sup\{\pi(w) | \|A\|_w \geq i\}$$

(possibility of A having value $\geq i$). σ_A is a fuzzy truth value; clearly, we want

$$\|A \lhd B\| = 1 \text{ iff } (\forall i)(\sigma_A(i) \leq \sigma_B(i)),$$

thus we define

$$\|A \lhd B\| = \min_i \max_j \{I(i,j) | \Pi(\|A\| = i) \leq \Pi(\|B\| = j)\}$$

This is another comparative fuzzy possibilistic logic CFPL'. Translate CFPL'-formulas into MTL-formulas by defining

$$(A \lhd B)^* = \Box(A \rightarrow \hat{F}B)$$

($\Box C$ being $\hat{G}C \wedge \hat{H}C$).

Theorem: The following are equivalent:
(1) A has a CFPL'-model,
(2) A has a finite CFPL'-model,
(3) A^* has a MTL-model,
(4) A^* has a finite MTL-model.

Thus $*$ is a faithful embedding of CFPL' into MTL.

4. CONCLUSION.

In fact, the paper is a kind of *postscript* to Part I; we have referred on new results obtained since Part I was written. In this way we hope to have given the reader an up-to-date picture of formal logical systems of approximate reasoning. Several problems remain, in particular, to characterize the following:

Some finite-valued comparative fuzzy uncertainty logics, as:
— belief function logic,
— probabilistic logic.

Some infinite-valued fuzzy logics, among them fuzzy version of:
— S5, KD45, tense
— comparative possibilistic, probabilistic, belief function logic.

Nevertheless, it appears to have been shown that natural modal systems of logics of uncertainty can be formulated, axiomatized and related to some classical systems of modal logic and their fuzzifications.

REFERENCES:

1. Hájek, P.: On logics of approximate reasoning, Neural Network Word, 6 (1993), 733–744.

2. Hájek, P.: On logics of approximate reasoning, in: Knowledge Representation and Reasoning Under Uncertainty (1994), M. Masuch and L. Pólos, Eds., Springer-Verlag, Heidelberg, pp. 17–29.

3. Hájek, P.: Fuzzy logic and arithmetical hierarchy, Fuzzy Sets and Systems (to appear).

4. Hájek, P.: Fuzzy logic as logic, in: Mathematical Models of Handling Partial Knowledge in Artificial Intelligence (Erice (Italy), 1994), G. Coletti and all, Eds., Pergamon Press.

5. Voorbraak, F.: As far as i know. Epistemic logic and uncertainty, Department of Philosophy Utrecht University, disertation, 1993.

6. Shafer, G.: Allocations of probability, The Annals of Probability 7, 5 (1979), 827–839.

7. Harmanec, D., Klir, G. J., and Resconi, G.: On modal logic interpretation of Dempster-Shafer theory of evidence, International Journal of Intelligent Systems 9, 10 (1994), 941–951.

8. Hájek, P.: Getting belief functions from Kripke models, International Journal of General Systems (to appear).

9. Pearl, J.: Probabilistic Reasoning in Intelligent Systems: Networks of Plausible Inference, Morgan Kaufmann Publishers, San Mateo, California, 1988.

10. Smets, P.: Probability of provability and belief functions, Logique et Analyse 133-134 (1991), 177–195.

11. Bendová, K., and Hájek, P.: Possibilistic logic as a tense logic, in: Qualitative Reasoning and Decision Technologies (Proceedings of QUARDET'93) (Barcelona, 1993), N. P. Carreté et al., Eds., CIMNE, pp. 441–450.

12. Harmanec, D., and Hájek, P.: A qualitative belief logics, International Journal of Uncertainty, Fuzziness and Knowledge-Based Systems 2, 2 (1994), 227–236.

13. Hájek, P., Harmancová, D., Esteva, F., Garcia, P., and Godo, L.: On modal logics for qualitative possibility in a fuzzy setting, in: Uncertainty in Artificial Intelligence; Proceedings of the Tenth Conference (Seattle, WA, 1994), R. López de Mántaras and D. Poole, Eds.

14. Hájek, P., Harmancová, D., and Verbrugge, R.: A qualitative fuzzy possibilistic logic, International Journal of Approximate Reasoning (to appear).

BASED REVISION OPERATIONS AND SCHEMES: SEMANTICS, REPRESENTATION, AND COMPLEXITY

B. Nebel

University of Ulm, Ulm, Germany

Abstract

The theory of belief revision developed by Gärdenfors and his colleagues characterizes the classes of reasonable belief revision operations. However, some of the assumptions made in the theory of belief revision are unrealistic from a computational point of view. We address this problem by considering revision operations that are based on a priority ordering over a set of sentences representing a belief state instead of using preference relations over all sentences that are accepted in a belief state. In addition to providing a semantic justification for such operations, we investigate also the computational complexity. We show how to generate an epistemic entrenchment ordering for a belief state from an arbitrary priority ordering over a set of sentences representing the belief state and show that the resulting revision is very efficient. Finally, we show that some schemes for generating revision operations from bases can encode the preference relations more concisely than others.

1   Introduction

The problem of changing a belief state in the face of new information arises in a number of areas in computer science and artificial intelligence. e.g., in updating logical database. in hypothetical reasoning. etc. Most of the research in this area is influenced by work in philosophical logic, in particular by Gärdenfors and his colleagues [1. 2. 3]. who developed the *theory of belief revision* (see also [4].) The

---

*This work was supported by the German Ministry for Research and Technology (BMFT) under grant ITW 8901 8 as part of the WIP project and by the European Commission as part of DRUMS-II. the ESPRIT Basic Research Project P6156.

related problem of how to accommodate new information that is the result of a change in the world has been analyzed by Katsuno and Mendelzon [5], who call the corresponding belief change operations *belief updates* in order to distinguish them from *belief revision.*

The main topic of the work by Gärdenfors and his colleagues is the characterization of *classes* of reasonable *revision operations.* where belief states are modeled by *deductively closed sets of propositional sentences* (so-called *belief sets*) [2] or. equivalently, as a set of models [6]. Starting with a number of *rationality postulates*, the set of possible change operations on a belief state is constrained to those that fulfill the postulates. Based on that, it is possible to define and analyze specific *revision schemes* that generate revision operations by employing some *preference* information. One such revision scheme is. for instance. the *partial meet revision* scheme [1]. Another scheme uses so-called *epistemic entrenchment orderings* in order to generate revision operations [7].

If one wants to apply this theory in a computer science or artificial intelligence application. there are two severe problems. First of all. the assumption that belief states are modelled by deductively closed sets of sentences seems to be computationally infeasible. However, we can, of course, represent belief sets by a finite *belief base*, a finite set of sentences that is logically equivalent with the belief set (provided the belief set is finite modulo logical equivalence). Secondly. there is the problem that the extra information required by the above mentioned revision schemes is usually a relation over the set of all sentences in a deductively closed theory, which is representationally infeasible.

One way to address the two problems is to consider revision operations on belief bases (so-called *base revisions*), i.e., operations that modify a belief base instead of a belief set [8. 9. 10, 11, 12, 13]. Such an approach often also matches certain characteristics of an application setting very well. If. for instance, a code of norms or a scientific or naive theory of the world is represented by a set of explicitly stated sentences. one may want to express preferences between these sentences.

However, the base revision approach violates the principle of "irrelevance of syntax" [14, 6], i.e., base revisions may lead to different results for belief bases that are syntactically different but logically equivalent. We address this criticism by proposing a different view on base revisions. Instead of analyzing them on the base level, we consider equivalent belief revision operations that are *generated* by the preferences on the base.

We focus on revision schemes that are as efficient as possible. namely. those that generate revision operations that are computable using a polynomial number of NP-oracle calls and which allow a concise representation of the revised base.

2 Preliminaries

Throughout this paper, a finitary propositional language $\mathcal{L}$ with the usual logical connectives ($\neg$, $\vee$, $\wedge$, $\rightarrow$ and $\leftrightarrow$) is assumed. The finite alphabet of propositional variables $p, q, r \ldots$ is denoted by Σ, propositional sentences by $\tau, \omega, \phi, \psi, \chi, \ldots$, constant truth by $\top$, its negation by $\bot$, and sets of propositional sentences by $K, L, M, \ldots$ and $A, B, C, \ldots$

The symbol $\vdash$ denotes derivability and Cn the corresponding closure operation, i.e., $Cn(K) \overset{\text{def}}{=} \{\varphi \in \mathcal{L} \mid K \vdash \phi\}$. Instead of $Cn(\{\phi\})$, we will also write $Cn(\phi)$. Deductively closed sets of propositional sentences, i.e., $K = Cn(K)$, are denoted by the capital letters $K, L, M \ldots$ and are called **belief sets**. The set of all belief sets over $\mathcal{L}$ is denoted by $Th_{\mathcal{L}}$. The monotonic addition of a propositional sentence ϕ to a belief set K, i.e., $Cn(K \cup \{\varphi\})$, is denoted by $K + \phi$ and called **expansion** of K by ϕ. Arbitrary sets of sentences are called **belief bases** and are denoted by capital letters from the beginning of the alphabet. Systems of belief bases and belief sets are denoted by S. Finite belief bases C are often identified with the conjunction of all propositions $\bigwedge C$. If $S = \{A_1, \ldots, A_n\}$ is a finite family of finite belief bases, then $\bigvee S$ shall denote a proposition logically equivalent to $(\bigwedge A_1) \vee \ldots \vee (\bigwedge A_n)$. As usual, we set $\bigvee \emptyset = \bot$.

A **belief revision operation** is a function [1, 2]:

$$\dot{+}: Th_{\mathcal{L}} \times \mathcal{L} \rightarrow Th_{\mathcal{L}}. \tag{1}$$

where the result of the operation is denoted by $K \dot{+} \omega$, which is intended to be a consistent belief set that differs *minimally* from K and contains ω. While these conditions do not lead to a unique operation, it is possible to constrain the space of reasonable belief revision operations. Gärdenfors proposed the following set of **rationality postulates for revision operations**:

($\dot{+}$1) $K \dot{+} \omega$ is a belief set;

($\dot{+}$2) $\omega \in K \dot{+} \omega$;

($\dot{+}$3) $K \dot{+} \omega \subseteq K + \omega$;

($\dot{+}$4) If $\neg \omega \notin K$, then $K + \omega \subseteq K \dot{+} \omega$;

($\dot{+}$5) $K \dot{+} \omega = Cn(\bot)$ only if $\vdash \neg \omega$;

($\dot{+}$6) If $\vdash \omega \leftrightarrow v$ then $K \dot{+} \omega = K \dot{+} v$;

($\dot{+}$7) $K \dot{+} (\omega \wedge v) \subseteq (K \dot{+} \omega) + v$;

($\dot{+}$8) If $\neg v \notin K \dot{+} \omega$, then $(K \dot{+} \omega) + v \subseteq K \dot{+} (\omega \wedge v)$.

These postulates intend to capture the intuitive meaning of minimal change from a logical point of view [1, 2, 15, 4]. The first six postulates, which are straightforward, are called **basic postulates**. The two last postulates, which are called **supplementary postulates**, are less obvious. They capture the idea that a revision of K by a conjunction $(\phi \wedge \psi)$ should be achieved through a revision by ϕ and an expansion by ψ, if this is possible at all, i.e., if ψ is consistent with $K \dot{+} \phi$.

One interesting point to note is that the postulates do not constrain the revision operation with respect to varying belief sets. In other words, we may restrict ourselves to a given belief set and consider the mapping from $\mathcal{L}$ (the new information) to $Th_{\mathcal{L}}$ (the revised belief set).

Based on the above framework, one can consider different *schemes* to generate revision operations. Formally, a **revision scheme** maps a belief set and some extra information, which encodes preferences, to a belief revision operation on the given belief set. In our setting (of finitary propositional logic), such a scheme can be considered as an *algorithm* that computes the revision.

In [1], so-called partial meet revisions are investigated. This notion is based on systems of maximal (w.r.t. to set-inclusion) subsets of a given belief set K that do not allow the derivation of ϕ, called the **removal** of ϕ and written $K \downarrow \phi$:

$$K \downarrow \phi \stackrel{\text{def}}{=} \{L \subseteq K \mid L \not\vdash \phi, \forall M: L \subset M \subseteq K \Rightarrow M \vdash \phi\}. \qquad (2)$$

A **partial meet revision** (on K for all ϕ) is defined by a *selection function* γ that selects a nonempty subset of $K \downarrow \neg\phi$ (provided $K \downarrow \neg\phi$ is nonempty, $\emptyset$ otherwise) in the following way:

$$K \dot{+} \phi \stackrel{\text{def}}{=} \left(\bigcap \gamma(K \downarrow \neg\phi)\right) + \phi. \qquad (3)$$

Such partial meet revisions satisfy unconditionally the first six postulates. Furthermore, it is possible to show that all revision operations satisfying the basic postulates are partial meet revisions [1, Observation 2.5].

Instead of providing the preference information by a selection function, one may also think of preference relations over sentences. **Epistemic entrenchment orderings**, written as $\phi \preceq \psi$, are defined over the entire set of sentences $\mathcal{L}$ and have to satisfy the following properties:

($\preceq$1) If $\phi \preceq \psi$ and $\psi \preceq \chi$, then $\phi \preceq \chi$.

($\preceq$2) If $\phi \vdash \psi$, then $\phi \preceq \psi$.

($\preceq$3) For any ϕ, ψ, $\phi \preceq (\phi \wedge \psi)$ or $\psi \preceq (\phi \wedge \psi)$.

($\preceq$4) When $K \neq Cn(\bot)$, then $\phi \notin K$ iff $\phi \preceq \psi$ for all $\psi \in \mathcal{L}$.

($\preceq$5) If $\psi \preceq \phi$ for all $\psi \in \mathcal{L}$, then $\vdash \phi$.

From ($\preceq$1) and ($\preceq$3), it follows that an epistemic entrenchment ordering is a complete preorder over $\mathcal{L}$. The strict part of this preorder will be denoted by $\prec$ in the following.

Using such a relation, one can define a revision scheme. which we will call **cut revision**:

$$K \dotplus \phi \overset{\text{def}}{=} \{v \in K \mid \neg o \prec v\} + o. \tag{4}$$

From results by Gärdenfors and Makinson [7] and Rott [16]. it follows that class of belief revision operations generated by this scheme coincides with the class of revision operations satisfying all rationality postulates.

3 Prioritized Meet Base-Revision

Although the theory sketched above provides us with a good picture of the ways a belief set can be revised, it does seem not to be possible to use this theory for implementing belief revision on a computer system. First of all. the assumption that a belief state is modeled as a deductively closed set of sentences sounds unrealistic. Secondly. the amount of preference information needed for the partial meet revision and the epistemic entrenchment scheme seems to be prohibitive. e.g. as has been shown by Gärdenfors and Makinson. one needs an ordering over $2^{|\Sigma|}$ sentences to specify an entrenchment ordering.

Addressing these problems. we consider revision schemes that use preference information that has a size polynomial in the size of a belief base. In particular. we focus on schemes where the preference information is encoded by a *complete preorder* $\trianglelefteq$ over the set of sentences in a belief base. also called **epistemic relevance ordering** [17, 12], with the intuitive meaning that if $o \trianglelefteq v$. then v is at least as relevant. important, or reliable than o (see also. e.g.. [18, 8, 19]). Equivalently. we can view a base A as partitioned into n **priority classes** or *ranks* $A_1, \ldots, A_n$ as follows (using $\triangleleft$ to denote the strict part of $\trianglelefteq$):

$$\phi \in A_1 \quad \text{iff} \quad \forall v \in A : \phi \trianglelefteq v.$$

$$\phi \in A_{i+1} \quad \text{iff} \quad \exists v \in A_i : \big(v \triangleleft \phi \wedge \neg(\exists \chi \in A : v \triangleleft \chi \triangleleft \phi)\big).$$

The union of the highest priority classes down to the jth class. i.e., $\bigcup_{i=j}^{n} A_i$. will also be written as $\overline{A_j}$.

Based on a such an epistemic relevance ordering. one can define a removal operation . called **prioritized base-removal** and written $A \Downarrow o$. that keeps as many sentences with high priority as possible:

$$A \Downarrow \phi \overset{\text{def}}{=} \{B \subseteq A \mid B \nvdash \phi, \forall C. j : B \cap \overline{A_j} \subset C \cap \overline{A_j} \tag{5}$$
$$\Rightarrow C \cap \overline{A_j} \vdash \phi\}.$$

Similar to partial meet revision. we define **prioritized meet base-revision**. written $A \Diamond \phi$, on a prioritized base A as follows:

$$A \Diamond \phi \overset{\text{def}}{=} \Big(\bigcap_{B \in (A \Downarrow \neg \phi)} Cn(B)\Big) + \phi. \tag{6}$$

As mentioned above, we will view $\dot{+}$ as a base-revision *scheme*, i.e., as defining a belief-revision operation $\dot{+}$ on $Cn(A)$ using A and $\trianglelefteq$ as "parameters:"

$$Cn(A) \dot{+} \phi \stackrel{\mathrm{def}}{=} A \dot{+} \phi. \qquad (7)$$

Although the construction looks quite plausible, it leads to a number of problems. First of all, revision operations generated by the prioritized meet base-revision scheme do not satisfy all rationality postulates [12]. Secondly, there is the problem of constructing a representation of the result of the revision operation. In our case of belief sets that are finite modulo logical equivalence, such a base can be easily specified. If A is a prioritized belief base then

$$A \dot{+} \phi = Cn\Big((\bigvee(A \Downarrow \neg\phi)) \wedge \phi\Big). \qquad (8)$$

However, the resulting base looks quite unintuitive. Furthermore, the cardinality of $(A \Downarrow \phi)$ cannot be polynomially bounded as the following example demonstrates. Let

$$A = \{p_1, \ldots, p_m, q_1, \ldots, q_m\} \qquad (9)$$

$$\phi = \bigwedge_{i=1}^{m} (p_i \leftrightarrow \neg q_i). \qquad (10)$$

and assume that there is just one priority class. Clearly, $(A \Downarrow \neg\phi)$ has exponentially many elements.[1]

Thirdly, prioritized meet base-revision is computational very difficult. Determining whether a proposition follows from a revised prioritized base is Π_2^p-complete[2] [12, 21].

The representational problem mentioned above could be avoided, if we defined the revision by considering the intersection over all maximal consistent sub-bases instead of using the intersection over the *consequences* of the maximal consistent sub-bases—a method that has been called *when in doubt, throw it out* (WIDTIO) [22]. However, the computational problems would remain. Deciding whether a proposition follows from a WIDTIO-revision is also Π_2^p-hard [21].

4   Cut Base-Revision

As mentioned above, prioritized meet base-revision suffers from a number of problems. Accounting for these problems, we will consider a base-revision scheme that resembles the *cut revision scheme* on belief sets (see Eq. (1)).

---

[1] In fact, from a recent result by Cadoli, Donini, and Schaerf (personal communication), it follows that in the general case it is impossible to find a dense (i.e., polynomial) representation of a revised base unless the polynomial hierarchy collapses.

[2] It is assumed that the reader is familiar with basic notions of complexity theory (see, e.g., [20]).

As before, we assume that the base is partitioned into priority classes $A_1 \ldots \ldots A_n$. The **cut base-revision**, written $A \otimes \phi$, is then defined as follows:

$$A \otimes \phi \stackrel{\text{def}}{=} Cn(\{v \in A | v \in A_j, \overline{A_j} \not\vdash \neg o\}) + o. \qquad (11)$$

Viewing $\otimes$ as a revision scheme, one easily verifies that all basic revision postulates are satisfied.

Proposition 1 *Belief revision operations generated by the cut base-revision scheme satisfy the basic rationality postulates.*

Moreover, contrary to the prioritized meet base-revision scheme, the result of a revision can be straightforwardly represented as a belief base, namely, as $\overline{A_j} \cup \{o\}$ with j being the smallest index such that $\overline{A_j}$ is consistent with $\{o\}$. Besides the representational advantages, cut base-revisions are also better behaved from a computational point of view.

Proposition 2 *Deciding $A \otimes \phi \vdash \psi$ is in $P^{NP[O(\log n)]}$.*

Since belief revision in general is a problem that is NP-hard and co-NP-hard [12], the above result is close to the optimum. Further, if we reduce the complexity of propositional reasoning, for instance, by restricting the expressiveness, we can obtain a polynomial-time revision scheme.

While all the above sounds as if cut base-revisions are much more well-behaved than prioritized meet base-revisions, there are also apparent disadvantages. Firstly, cut base-revision is much more radical in giving up beliefs than prioritized meet base-revision (or even WIDTIO-revisions). It cuts away all priority classes that have a priority equal or lower to the class that is "responsible" for an inconsistency. For this reason, one might argue that this kind of revision violates the principle of *minimal modification*. Secondly, it is not clear whether cut base-revision satisfies the supplementary postulates as well. In order to address these problems, we will investigate the relationship between epistemic relevance and epistemic entrenchment.

5 Relevance vs. Entrenchment in Cut Base-Revision

Epistemic relevance differs from epistemic entrenchment in two aspects. Firstly, the former is a complete preorder over a belief base while the second is a complete preorder over $\mathcal{L}$. Secondly, epistemic entrenchment respects the logical contents of the sentences while epistemic relevance is an arbitrary preorder. For instance, we may well have the case that $\phi \vdash \psi$ but $\psi \lhd \phi$, contradicting ($\preceq 2$).

In the example above. it does not seem to make much sense to give ϕ higher priority than v since o has to be retracted in any case if v is forced to be deleted. More generally. if the sub-base $C \subseteq A$ implies $\phi \in A$, then it does not make much sense that o has a priority that is lower than the minimum of the priorities of the sentences in C. For this reason, let us assume that the epistemic relevance ordering satisfies the following **priority consistency condition** (PCC) (see also [23. 24. 25]):

> For all $o \in A$. if C is a nonempty subset of A such that $C \vdash o$. then there exists $\chi \in C$ such that $\chi \unlhd \phi$.

As has been shown by Rott [24]. this condition is necessary and sufficient for the **extendibility** of $\unlhd$ on A to an epistemic entrenchment ordering. i.e.. an epistemic entrenchment ordering $\preceq$ satisfying for all $\phi, v \in A$ that $o \unlhd v$ iff $o \preceq v$.

Rott [24] gives the following construction to show sufficiency (adapted to finite bases). Let A be a prioritized base that satisfies (PCC) and assume w.l.g. that $A_n = \emptyset$. Then define $\preceq$ on $\mathcal{L}$ by

$$o \preceq v \quad \text{iff} \quad A \nvdash o \text{ or } \max_j(\overline{A_j} \vdash o) \leq \max_i(\overline{A_i} \vdash v) \tag{12}$$

This construction does not only show that a priority consistent epistemic relevance ordering $\unlhd$ can be extended to an entrenchment ordering $\preceq$. but also provides us with an entrenchment ordering that **minimally extends** the relevance ordering, i.e.. if $\phi \prec v$. then all other entrenchments $\preceq'$ extending $\unlhd$ also contain $\phi \prec' v$.

Theorem 3 *Let A be a base with an associated epistemic relevance ordering $\unlhd$ satisfying (PCC). Then $\preceq$ determined by Eq. (12) is the unique minimal extension of $\unlhd$ to an epistemic entrenchment ordering over $Cn(A)$.*

Even more interestingly. the cut base-revision operation on a priority consistent base coincides with the cut revision using the epistemic entrenchment ordering generated by Eq. (12).

Theorem 4 *Let A be a belief base and $\unlhd$ be an associated epistemic relevance ordering satisfying (PCC). Let $\preceq$ be the epistemic entrenchment order generated from $A, \unlhd$ according to Eq. (12) and $\dotplus$ be the cut revision on $Cn(A)$ using $\preceq$. Then*

$$Cn(A) \dotplus o = Cn(A - o). \tag{13}$$

In other words, priority-consistent epistemic relevance orderings can be viewed as a dense representation of an epistemic entrenchment ordering. Furthermore, assuming that belief sets are finite modulo logical equivalence, all belief revision operations generated by the cut revision (i.e., all fully rational revision operations) are generated by some priority-consistent epistemic relevance ordering. Hence, the cut base-revision scheme can generate all fully rational revision operations.

This leaves us with the question of what kind of belief revision operation is generated if the epistemic relevance ordering is not priority consistent. This question is important because it seems not very realistic to put the burden of guaranteeing this condition, which involves deciding propositional decidability, on the shoulders of a user.

As it turns out, we can interpret the specified priorities as an *approximation* specifying *lower bounds* for the intended priorities (see also [26, 27]). Using (12) on some arbitrary prioritized base A, the resulting relation $\preceq$ is again an epistemic entrenchment ordering for the generated belief set $Cn(A)$. Furthermore, $\preceq$ satisfies a number of conditions that show that $\preceq$ can be indeed regarded as the epistemic entrenchment relation intended by the priorization of A.

Theorem 5 *Let A be an arbitrary prioritized base. Then the ordering $\preceq$ generated by Eq. (12) is an epistemic entrenchment ordering for $Cn(A)$, and the restriction of $\preceq$ to A is a priority-consistent epistemic relevance ordering that generates $\preceq$ according to Eq. (12).*

Furthermore, the cut base-revision coincides with the cut revision using the generated epistemic entrenchment.

Theorem 6 *Let A be an arbitrary prioritized base and $\dotplus$ be a cut revision operation based on the epistemic entrenchment ordering generated by (12). Then*

$$Cn(A) \dotplus \phi = Cn(A \otimes \phi). \tag{14}$$

From that the following corollary follows straightforwardly.

Corollary 7 *The cut base-revision scheme coincides with the set of belief revision operations that satisfy all rationality postulates.*

6   Linear Base-Revision

While prioritized meet base-revision has a lot of conceptual, representational, and computational problems, cut base-revision seems to be too radical in deleting beliefs. In trying to find a compromise, one may consider the method of deleting an entire priority class if one sentence in it leads to a contradiction that cannot be blamed on sentences in lower priority classes. but keeping as many of the other priority classes as possible.[3] We could view this scheme as if we had prioritized bases where each priority class contains only one element. Formally, given a prioritized base A with n priority classes. we form a new prioritized base with n classes where $B_i = \{\bigwedge A_i\}$. The resulting epistemic relevance ordering on B is then a linear order.

Interestingly. the prioritized meet base-revision scheme on linearly ordered bases behaves in the way as spelled out above. i.e., it deletes a priority class if it is to be blamed for a contradiction and no lower class can be blamed for it. Since prioritized meet base-revision on linearly ordered bases seems to be an important special case. we will consider it as a base-revision scheme on its own. as the **linear base-revision scheme**. The **linear base-revision operation** will be denoted by $\odot$.

As can be easily shown. the linear base-revision operation picks just one subset of the base as the result of the revision [12, Proposition 7] (see also [11]). Furthermore. the linear base-revisions scheme satisfies all rationality postulates [12, Theorem 8]. Finally, also the computational properties of this scheme are very attractive.

Theorem 8 *The problem of deciding $A \odot \vdash \varphi$ is $\mathrm{P}^{\mathrm{NP}[O(n)]}$-complete.*

One interesting question is how linear base-revision relates to cut base-revision. As can be easily shown. cut base-revisions can be polynomially transformed to linear base-revisions.

Proposition 9 *Let A be a prioritized base. Then there exists a linearly prioritized base B that can be computed in polynomial time such that*

$$Cn(A \odot \varphi) = Cn(B \odot \varphi). \tag{15}$$

Using this result and [12, Theorem 8]. the following corollary is immediate.

Corollary 10 *The linear base-revision scheme coincides with the set of belief revision operations satisfying all rationality postulates.*

Although this result shows that everything that can be expressed as a cut base-revision can be expressed as a linear base-revision (and *vice versa*). it does not imply that the representation of the bases are equally concise. In fact, the complexity results show that this can only be the case if $\mathrm{P}^{\mathrm{NP}[O(\log^2 n)]} = \mathrm{P}^{\mathrm{NP}[O(n)]}$. which is considered to be rather unlikely. In fact, the most natural transformation of linear base-revision to cut base-revision blows up the belief base exponentially.

---

[3]This is also the intention of the revision of possibilistic knowledge bases described in [27. p. 167].

Assume a linearly ordered prioritized base A with n priority classes. Then we define a logically equivalent belief base $B = \pi(A)$ with $2^n - 1$ priority classes. The priority classes of B are again singletons, and the elements of these classes are disjunctions of the classes in A. In particular, the sentence $A_{j_1} \vee \ldots \vee A_{j_k}$ is in the priority class B_l, where $l = \sum_{l=1}^{k} 2^{(j_l - 1)}$.

Theorem 11 *Let A be a prioritized base with a linear epistemic relevance ordering. Then it holds that*

$$Cn(A \odot \phi) = Cn(\pi(A) \otimes o). \qquad (16)$$

As mentioned above, it seems unlikely that we can find a transformation that is "cheaper," i.e., uses less than exponential time. Hence, it appears to be the case that the linear base-revision scheme uses a more concise coding of preference information than the cut base-revision scheme.

7   Concluding Remarks

Revision schemes that generate a belief revision operation from a given belief base and some additional preference information that is not of infeasible size seem to be of practical interest in the area of belief revision. The most straightforward such scheme, *prioritized meet base-revision*, has a number of conceptual and computational problems, though. Adopting the notion of *epistemic entrenchment*, we were able to show that priorities of sentences in belief bases can be interpreted naturally as lower bounds of epistemic entrenchment, and based on this view, it is possible to define an elegant and efficient revision scheme, called *cut base-revision* scheme. Relating this scheme to earlier results concerning prioritized meet base-revision on linearly ordered prioritized bases, we noted that the latter is expressively equivalent to the former. Furthermore, we noted that it appears to be the case that prioritized meet base revision on linearly ordered bases permits to state the preference information in a way that is more concise than in the case of cut base revisions.

Acknowledgements

I would like to thank Didier Dubois, Dov Gabbay, and Henri Prade for discussion on the subject of this paper, Georg Gottlob and Werner Nutt for some comments on computational complexity issues, and Hans Rott for comments on an earlier version of this paper.

References

[1] Carlos E. Alchourrón. Peter Gärdenfors, and David Makinson. On the logic of theory change: Partial meet contraction and revision functions. *Journal of Symbolic Logic*. 50(2):510–530, June 1985.

[2] Peter Gärdenfors. *Knowledge in Flux—Modeling the Dynamics of Epistemic States*. MIT Press, Cambridge, MA. 1988.

[3] P. Gärdenfors, editor. *Belief Revision*, volume 29 of *Cambridge Tracts in Theoretical Computer Science*. Cambridge University Press, Cambridge. UK. 1992.

[4] Peter Gärdenfors and Hans Rott. Belief revision. Lund University Cognitive Studies 11. Cognitive Science. Department of Philosophy. University of Lund, Lund. Sweden. July 1992. To appear in: D. Gabbay (ed.), *Handbook of Logic in AI and Logic Programming*. Vol. IV: Epistemic and Temporal Reasoning.

[5] Hirofumi Katsuno and Alberto O. Mendelzon. On the difference between updating a knowledge base and revising it. In Gärdenfors [3]. pages 183–203.

[6] Hirofumi Katsuno and Alberto O. Mendelzon. Propositional knowledge base revision and minimal change. *Artificial Intelligence*. 52:263–294. 1991.

[7] Peter Gärdenfors and David Makinson. Revision of knowledge systems using epistemic entrenchment. In *Theoretical Aspects of Reasoning about Knowledge: Proceedings of the Second Conference*. Morgan Kaufmann, Asilomar. CA, 1988.

[8] Ronald Fagin, Jeffrey D. Ullman, and Moshe Y. Vardi. On the semantics of updates in databases. In *2nd ACM SIGACT-SIGMOD Symposium on Principles of Database Systems*. pages 352–365. Atlanta. Ga.. 1983.

[9] André Fuhrmann. Theory contraction through base contraction. *Journal of Philosophical Logic*. 20:175–203. 1991.

[10] Sven O. Hansson. *Belief base dynamics*. doctoral dissertation. Uppsala University, Sweden, 1991.

[11] Bernhard Nebel. A knowledge level analysis of belief revision. In R. Brachman, H. J. Levesque, and R. Reiter. editors. *Principles of Knowledge Representation and Reasoning: Proceedings of the 1st International Conference*. pages 301–311. Toronto. ON. May 1989. Morgan Kaufmann.

[12] Bernhard Nebel. Belief revision and default reasoning: Syntax-based approaches. In J. A. Allen, R. Fikes, and E. Sandewall, editors, *Principles of Knowledge Representation and Reasoning: Proceedings of the 2nd International Conference*, pages 417–428, Cambridge, MA, April 1991. Morgan Kaufmann.

[13] Hans Rott. Belief contraction in the context of the general theory of rational choice. *Journal of Symbolic Logic*, 58, December 1993.

[14] Mukesh Dalal. Investigations into a theory of knowledge base revision: Preliminary report. In AAAI-88 [28], pages 475–479.

[15] Peter Gärdenfors. Belief revision: An introduction. In Gärdenfors [3], pages 1–28.

[16] Hans Rott. Two methods of constructing contractions and revisions of knowledge systems. *Journal of Philosophical Logic*, 20:149–173, 1991.

[17] Bernhard Nebel. *Reasoning and Revision in Hybrid Representation Systems*, volume 422 of *Lecture Notes in Artificial Intelligence*. Springer-Verlag, Berlin, Heidelberg, New York, 1990.

[18] Gerhard Brewka. Preferred subtheories: An extended logical framework for default reasoning. In *Proceedings of the 11th International Joint Conference on Artificial Intelligence*, pages 1043–1048, Detroit, MI, August 1989. Morgan Kaufmann.

[19] Matthew L. Ginsberg. Counterfactuals. *Artificial Intelligence*, 30(1):35–79, October 1986.

[20] David S. Johnson. A catalog of complexity classes. In J. van Leeuwen, editor, *Handbook of Theoretical Computer Science, Vol. A*, pages 67–161. MIT Press, 1990.

[21] Thomas Eiter and Georg Gottlob. On the complexity of propositional knowledge base revision, updates, and counterfactuals. *Artificial Intelligence*, 57:227–270, 1992.

[22] Marianne S. Winslett. Reasoning about action using a possible models approach. In AAAI-88 [28], pages 89–93.

[23] N. Rescher. *The Coherence Theory of Truth*. Oxford University Press, Oxford, UK, 1973.

[24] Hans Rott. A nonmonotonic conditional logic for belief revision I. In A. Fuhrmann and M. Morreau, editors. *The Logic of Theory Change*, volume 465
of *Lecture Notes in Artificial Intelligence*, pages 135-183. Springer-Verlag,
Berlin, Heidelberg, New York, 1991.

[25] Mary-Anne Williams. *Transmutations of Knowledge Systems*. PhD thesis,
University of Sydney, Australia, 1994.

[26] Didier Dubois and Henri Prade. Epistemic entrenchment and possibilistic
logic. *Artificial Intelligence*, 50:223-239, 1991.

[27] Didier Dubois and Henri Prade. Belief change and possibility theory. In
Gärdenfors [3], pages 142-182.

[28] *Proceedings of the 7th National Conference of the American Association for
Artificial Intelligence*, Saint Paul, MI, August 1988.

ON THE NOTION OF SIMILARITY IN CASE-BASED REASONING

M.M. Richter

University of Kaiserslautern, Kaiserslautern, Germany

ABSTRACT

The semantics of similarity measures is studied and reduced to the evidence theory of Dempster and Shafer. Applications are given for classification and configuration, the latter uses utility theory in addition.

1. INTRODUCTION

Case-Based Reasoning (CBR) is one of the areas in Artificial Intelligence which is maturing to a technology (cf. e.g.[12]). A widely accepted technology depends usually on factors of heterogeneous character including a solid theoretical foundation, a framework of appropiate terminology, well-developed engineering methods, and a large experience in practical applications. In this article, we will contribute to the clarification of the concept of similarity which is crucial for CBR. For this purpose, we will first repeat some basic facts about CBR.

CBR can deal in principle with almost unlimited types of problems. If such a type is chosen a case is a pair (P, S) where P is a problem of this type and S is a solution for P; a case base CB is a set of such cases. We assume that (P, S_1) and (P, S_2) implies $S_1 = S_2$. This means that the solutions depend functionally on the problems and allows us to identify the cases with the problems and to include situations where we have problems and no solutions.
As examples, we consider two problem classes (cf. e.g. [11], [6]):

a) Analytic problems as the classification of objects;
b) Synthetic problems as the configuration of technical devices or the design of plans.

The usual description of how CBR proceeds is:

1) Present an actual problem P_a.
2) Select a case (P, S) from the case base CB such that P and P_a are "similar".
3) Transform the solution S for P into a solution S_a for P_a.

This simple description contains already the most important aspects
of CBR:

- the size and the structure of the case base
- the notion of similarity
- the retrieval problem for cases
- the notion of a solution transformation

One of the major difficulties, in particular for classification, comes from the fact that very often the problem is only partially described, i.e., we encounter a situation of incomplete information. In addition, the description may be noisy or uncertain.

For classification the solution transformation is often the identity which we will in the beginning assume here. Then the problem solving knowledge is contained in the case base and the specific similarity concept. The latter is usually given as a real valued function in order to express degrees of similarity. The literature is full of examples of such similarity measures and each running CBR system contains necessarily at least one measure. Sometimes the measure is not fixed and can be improved by a learning process, cf. e.g. the measure in PATDEX/2 (see [11]). The selection of a "similar" case from case base using a similarity measure μ is performed by applying the following principle:

Nearest Neighbor Principle NNP:

Given an actual problem P_a select a case (P,S) from CB such that $\mu(P_a, P)$ is minimal.

Notation: $P = NN(P_a, CB, \mu)$.

If the nearest neighbor is not uniquely defined then some selection procedure has to take place. In the sequel we will not consider such situations.

Although NNP is generally applied, it does not have a theoretical foundation like the Maximum Likelihood Principle in probability theory. In fact, for arbitrary measures μ other principles than just the nearest neighbor principle may be much more suitable.

Sometimes specific measures have a motivation coming from the problem situation, but in general the justification is just that it works quite well. To our knowledge, an attempt to give a formal semantics to similarity measures which justifies NNP has not been made. We will present an approach for specifiying such a semantics (or rather a "meaning") for similarity measures. This will result among others in an ideal measure reflecting precisely the available information when the selection has to take place. It is, however, not claimed that this is the only nor even the best approach; we rather hope to start a discussion on this topic.

2. SOME CONCEPTS FROM LOGIC

2.1 Classical Predicate Logic.

Next we will introduce some notions from logic in a way which is appropriate for our purposes.

We consider a class $\mathfrak{M}$ of structures M of predicate logic,

$$M = < U, (R_i)_{i \in I}, (f_j)_{j \in J} >$$

where U is the universe of of M, each R_i, is an n_i-place relation over U and each f_j is a partial n_j-ary function over U. Although neither U nor I or J are assumed to be fixed in $\mathfrak{M}$, we require that there are U_0, I_0 and J_0 such that for all structures in $\mathfrak{M}$

$$U_0 \subseteq U, I_0 \subseteq I \text{ and } J_0 \subseteq J$$

holds. This means that each model M has a reduct of the form

$$M_0 = <U_0, (R_i)_{i \in I_0}, (f_j)_{j \in J_0} >$$

M_0 is called the nucleus of M.

In most of the intended applications the universe will be finite. The situations we want to investigate lead to the following kind of structures.

a) Classification:

$$M = < U, R >, R \subseteq U$$

This partitions U into the two sets R and U\R; using more predicates classification problems with n classes can be formulated. If the number relational of classes is a priori unknown, the signature needs to be extended in order to introduce more classes. Also R will usually be defined in terms of other relations and functions which again requires a larger signature.

b) Configuration:

M describes a (complex) technical device or a plan using relational and functional dependencies. Adding new parts to the device results in general in an extension of the universe U while adding new dependencies requires an extension of the signature.

For a) a problem is given by an element a of U and the corresponding solution is the determination of the class to which a belongs. For b) the problem is a set of conditions on the model and the solution is a description of a model which satisfies these conditions.

For each structure M, we denote the corresponding first order predicate language by L(M). L(M) is assumed to contain a constant for each a $\in$ U. Furthermore we define

$$L(\mathfrak{M}) = \bigcup (L(M) \mid M \in \mathfrak{M})$$

We emphasize, however, that partial functions are admitted, mainly in order to cover incomplete attribute-value descriptions. In general, we do not distinguish notationally between the symbols in L(M) and the corresponding objects in M.

2.2 Alternative Model Descriptions

In most of our intended applications, the models of $\mathfrak{M}$ are not presented in the terminology of predicate logic but in various other ways. We call the formalisms used for this purpose model description languages; they can be quite arbitrary.

Definition: A model description language LModD is given by

(i) A recursive set called *set of expressions*; this set will again be denoted by LModD.
(ii) A computable function Sem: LModD $\rightarrow \mathfrak{p}(\mathfrak{M})$ called a semantic function where $\mathfrak{p}$ denotes the power set.

For S ∈ LModD the set Sem(S) is called the set of models for S. Clearly Sem is a generalization of the usual semantics for predicate logic. The most important example for classification is:

The models are of the form (U, R), $R \subseteq U$

(a1) $LModD = \{ (CB, f) \mid CB \subseteq U, f: U \to CB, f \upharpoonright CB = id\}$

The letters CB stand for "Case Base". If we assume that $R \cap CB$ is known, then Sem can be defined by reducing R to $R \cap CB$ using f:

$a \in R :\Leftrightarrow f(a) \in R$.

If the function f is only partially defined, the model given by Sem(CB, f) is not uniquely defined, and we obtain a set of models by the semantics function.

(a2) $LModD = \{(CB, \mu) \mid CB \subseteq U, \mu: U \times CB \to \mathbb{R}^+\}$

where $\mathbb{R}^+$ denotes the nonnegative reals.

μ is called a similarity measure which may meet some additional requirements. μ defines a function $f_\mu \to CB$ using the nearest neighbor principle NNP which defines a semantics as in (a1):
   $f_\mu(a) := NN(a, CB, \mu)$

(b) $LModD = L(M)$;

Syntactically, we put $LModD = L(\mathfrak{M})$.

For the semantics let $\vdash_{\mathfrak{R}}$ be some deductive operator.

For $\varphi \in L(\mathfrak{M})$ we define Sem(φ) by
$(a_1,..., a_n) \in R_i \quad :\Leftrightarrow \quad \varphi \vdash_{\mathfrak{R}} R_i(a_1,..., a_n)$

and

$f(a_1,..., a_n) = b \quad :\Leftrightarrow \quad \varphi \vdash_{\mathfrak{R}} f(a_1,..., a_n) = b$

Here the left sides of the equivalences take place in M while the right sides denote provability in $L(\mathfrak{M})$ using $\vdash_{\mathfrak{R}}$. The operator $\vdash_{\mathfrak{R}}$ may, e.g., denote derivability from a certain rule system $\mathfrak{R}$. This operator may have access to a similarity measure μ; such operators can occur in the context of configuration.

3. SIMILARITY

3.1 Some similarity measures

In the sequel, we assume that our objects (i.e., the problems) $x \in U$ are given as real valued vectors $x = (x_1,..., x_n)$, $x_i \in D_i$ which, e.g., may result from an attribute-value representation; D_i is called the i-th domain. Some father of many similarity measures is the Hamming measure μ_H (for the simple case of values from $\{0,1\}$):

$$\mu_H (x, y) = n - \sum_{i=1}^{n} | x_i - y_i |$$

If the individual attributes are of different importance, weight functions are introduced:

$$\mu_H (x, y) = \sum_{i=1}^{n} g_i (x_i, y_i),$$

where the g_i are real valued functions. This covers also the cases of general real valued attributes and the presence of noise. A special case is where μ_H is a linear function; then we deal with weighted Hamming measures. The weights are real numbers g_i which are supposed to reflect the importance of the i-th attribute:

$$\mu_H (x, y) = \sum_{x_i = y_i} g_i$$

The Tversky-measure is much more general and of the form (see[10]):

$$\mu_T = \alpha \cdot f(A) - \beta \cdot f(B) - \gamma \cdot f(C)$$

where α, β, γ are real numbers, f a real-valued function and
$$A = \{i \mid x_i = y_i\}$$
$$B = \{i \mid x_i = 1, y_i = 0\}$$
$$C = \{i \mid x_i = 0, y_i = 1\}.$$

As indicated above, our objects $x \in U$ may only be partially described, i.e., the values of some x_i may be missing. This causes the serious problem to extend these measures appropriately. Additional problems arise if

noise is present

- the values x_i are uncertain
- the values x_i are not independent
- a priori knowledge about the x_i is available.

There is of course the desire that the similarity measure reflects these aspects. As indicated in the introduction, we need a semantic interpretation of the real numbers which are values of the measure.

3.2 Similarity and Truth

We will deal here with problems of classification. The most striking difference between a similarity measure m and a truth evaluation function is that the first is real valued while the second is $\{0,1\}$-valued. It needs a device like the nearest neighbor principle NNP in order to obtain a binary decision from μ. There are some natural questions which arise in this context.

Suppose $a \in U$ and $x, y \in CB \subseteq U$ where x is the nearest neighbor of a in CB:

- what motivates to determine class (a): = class(x) instead class(a) := class(y)?
- which information is contained in the numerical value $\mu(a,x)$?
- which information is contained in the numbers $\mu(a,x) - \mu(a,y)$ and $\mu(x,y)$?

A similarity does not only determine the nearest neighbor, it provides some additional services like:
- the elements of CB are arranged on an ordinal scale;
- this arrangement is attached with real numbers, i.e., we obtain really a cardinal scale.
We have to answer the question what the meaning of the scale is. All of the above questions are related to question what the values of μ have to do with an approximation of the truth of a statement about class(a). In other words, which precise information about the truth of the equation class(a) = class(x) is contained in the number $\mu(a,x)$? The answer to this question would assign a meaning, i.e., a semantics to μ.

3.3 Similarity and Evidence

To simplify the situation, we will assume that all attributes are independent and no a priori knowledge is present; then the only available information consists in the knowledge of some attribute values. Let $a \in U$ and $x \in CB$. If some a_i are already observed a first approach would be to define

$$\mu(a,x) = \text{Prob}((a, x) \mid \text{given observations})$$

It is, however, difficult to assign such a conditional probability in a satisfying manner if only a few attributes are observed.

A known attribute value a_i, however, is a piece of information which hints to the set

$X_i = \{ x \in CB \mid x_i = a_i \}.$

Following J. Kohlas (cf. [2], [3], [4]) this gives rise to a basic evidence measure m_i on CB, i.e., to a probability measure on the power set $\mathfrak{p}(CB)$ provided we can quantify this hint on X_i by a real number g_i, $0 \leq g_i \leq 1$.

We define m_i by putting $m_i(X_i) = g_i$, $m_i(CB) = 1-g_i$, i.e., some evidence goes to X_i and the rest is ignorance. Therefore, the measure m_i has only two focal sets (i.e. sets with positive measure) and because no other knowledge is available, we cannot distinguish between the elements of X_i.

If more attributes values are observed the evidence measures can be accumulated using Dempster's rule (because of our independence assumption).

In general, Dempster's rule says for $X \neq \emptyset$ (cf. [1], [8]):

$$m_1 \oplus m_2 (X) = \sum_{Y \cap Z = X} m_1(Y) \cdot m_2(Z) \cdot \frac{1}{1-K}$$

$$\text{with } K = \sum_{Y \cap Z = \emptyset} m_1(Y) \cdot m_2(Z)$$

$$\text{and } m_1 \oplus m_2 (\emptyset) = 0$$

For $K \neq 0$ we have conflicts, and for $K = 1$ the accumulation $m_1 \oplus m_2$ is undefined.

Now we introduce some additional notation:

Suppose $I = \{1,...,n\}$; assume $J \subseteq I$:

$X_J = \{x \in CB \mid x_i = a_i, i \in J\}$, $X_i = X_{\{i\}}$
$m_J = \oplus (m_i \mid i \in J)$, $m_i = m_{\{i\}}$

Note that after a series of observations the sets X_J are closed under intersections.

If $X_{J1} = X_{J2}$ for $J_1 \neq J_2$ we call it a multiplicity. Without multiplicities and conflicts, Dempster's rule simplifies and gives for $J' \subseteq J \subseteq I$

$$m_J (X_{J'}) = \prod_{i \in J'} g_i * \prod_{i \in \mathbb{N}'} (1-g_i)$$

$$= \sum_{J'' \subseteq J\backslash J'} \left(\prod_{i \in J} g_i\right) * (-1)^{|J''|} * \prod_{k \in J''} g_k$$

Also:

$$m_J (CB) = \prod_{i \in J\backslash J'}(1-g_i) = 1 - \sum_{J'' \subseteq J\backslash J'} (-1)^{|J''|} * \prod_{k \in J''} g_k$$

Some $x \in CB$ may be elements of several focal sets X. We now make the crucial assumption that each such membership contributes to the similarity of x and a according to the evidence measure of each X. This leeds to the following definition:

<u>Definition:</u> (i) $v_J (X) = \sum_{Y \supseteq X} m_J (Y)$, Y a focal set for m_J

(ii) $v_J(x) = v_J(X)$, X the minimal focal set containing $x \in U$ (which is uniquely defined).

(iii) $\mu_J^D (a,x) = v_J(x)$, where a is the actual case.

If noise is present, we can proceed as follows:

$X_i^{\varepsilon,\delta} = \{x \in CB \mid \varepsilon \leq |X_i - a_i| \leq \delta \}$,

$m_i^{\varepsilon,\delta} (X_i^{\varepsilon,\delta}) = g^{\varepsilon,\delta}$, $m_i^{\varepsilon,\delta} (CB) = 1 - \sum_{a,\delta} g_i^{\varepsilon,\delta}$

for $0 \leq \varepsilon < \delta \leq 1$; $g_i^{\varepsilon,\delta}$ are again real numbers.

If the source of the information for the attribute value a_i is unreliable then the g_i will also reflect this uncertainty. If more than one independent source confirms this value we can reflect this by the accumulation of evidences in the measure. We note that to our knowledge such situations have been neglected in CBR.

We call μ_J^D the Dempster (similarity) measure. If the attributes are not independent, the measure can also be defined, but it requires a more refined rule than Dempster' s rule (cf. e.g. [4]).

We obtain trivially $v_J(X_J) = 1$. We also have for $J' \subseteq J \subseteq I$ if no multiplicities and conflicts occur

$$v_J(X_{J'}) = \sum_{J'' \subseteq J'} (\prod_{i \in J''} g_i \prod_{i \in J \setminus J''} (1-g_i))$$

Now suppose that $X_J = \emptyset$ but $X_{J'} \neq \emptyset$ for $J' = J \setminus \{ i \}$, some $i \in I$.

Neglecting renormalisation, we obtain for a minimal focal set $X_{J'}$:

$$v_J(X_{J'}) = \prod_{i \in J \setminus J'} (1 - g_i)$$

If $\sum_{i \in J} g_i = 1$ this gives

$$v_J(X_{J'}) = 1 - \sum_{i \in J \setminus J'} g_i + \sum_{i,j \in J \setminus J'} g_i g_j - \sum_{i,j,k \in J \setminus J'} g_i g_j g_k + - \cdots$$

$$= \sum_{i \in J'} g_i + \text{ terms of higher order}$$

or

$$\mu_J^D(a,x) = \sum_{x_i = a_i} g_i + \text{ terms of higher order.}$$

This means that in this situation, the evidence measure coincides with a weighted Hamming measure up to a small error. Because the evidence measure is difficult to compute (cf. [5]) for the computational complexity of Dempster's rule), we obtain a good motivation for the use of Hamming measures from the viewpoint of efficiency.

If conflicts occur, a normalization has to take place, but this will not change the ordering of the neighbors of a and the cardinal scale is only changed by a constant factor.

If multiplicities are allowed, the situation is, however, not so easy.

We take an example:

$a = (1,1,1,1)$, $x = (1,1,0,0)$, $y = (0,1,1,0)$, $z = (0,0,1,1)$, $CB = \{x,y,z\}$,
$X_{12} = X_1 \cap CB = X_1 \cap X_2$

$I = J = \{1,...,4\}$

We obtain

$$v_I(X_{12}) = 1 + g_1 - g_3 - g_4 - g_1 (g_2 + g_3 + g_4) + g_3 g_4 + g_2 g_2 g_3 + g_2 g_2 g_4 - g_1 g_2 g_3 g_4$$

Our approach leads also to an indiscernilibity relation $\approx$ in the sense of the theory of rough sets:

$x \approx y \Leftrightarrow$ x and y are in the same focal sets.

If the available information is rich enough that the singletons are the only focal sets, then the evidence measure is a probability measure on CB. In such a situation, we have the desired formula mentioned at the beginning of 3.3.:

$sim(a, b) = Prob(class(a) = class(b) \mid$ given observations)

In summary, the evidence measure μ_J^D can be seen as the measure reflecting exactly the given information. For this measure the Nearest Neighbor Priniciple is clearly justified because it is nothing than the Maximum Likelihood Principle (applied to the evidence measure m which is a probability measure on the power set of the case base). This similarity measure may in concrete situations be difficult to compute or to approximate. However, there is now a reason to employ the results of probability theory and statistics for such purposes. In practice, this will result in the design of adaption algorithms for the measure.

3.4 Evidence and Utility

In this section, we will finally sketch some aspects of similarity in the context of configuration and planning.
The notion of truth applies only partially to configuration. A configuration may or may not be correct (i.e., meet some requirements), but it may also be more or less optimal with respect to some specified preferences. There, the truth value has to be replaced by a pair

(α, β)

where α is a value measuring correctness while β measures the degree of optimality. In addition, we have also the solution transformation T which means that we have to question the semantics of (μ, T) as

Semantics$(\mu, T) = (\alpha, \beta)$.

In order to consider a similarity measure μ, we will fix the transformation T for the rest of the paper. If we assume that T always checks for correctness, we have only to deal with the parameter β. This parameter is the form

$\beta = f(\beta_1, \beta_2)$

where β_1 measures the cost of T and β_2 measures the optimality of the solution. For the classification problems considered so far these costs were zero because T was the identity transformation. In the worst case, the case base contains no information and replanning takes place; then the costs are maximal.

Now another difference between classification and configuration enters the scenario. For classification, it is easy to check the correctness of the CB a posteriori. Therefore, CB contains only correctly classified cases, but in a case base for configuration, the cases usually will have suboptimal solutions. If a suboptimal solution is obtained from a similar case by applying T, this is not necessarily the result of an insufficient similarity or a bad transformation T, but may be entirely due to the fact that the solution of the case to which T was applied was not optimal. For technical reasons we therefore assume that all cases in CB have optimal solutions.

In the classical framework of utility theory one considers

- a set of situations $S = \{ S_i \mid i \in L \}$

- a set of actions $A = \{ A_k \mid k \in K \}$

- a set of real valued utilities $\{ \mu_{ik} \mid i \in L, k \in K \}$ which measure the utility of action A_k in situation S_i.

If a probability distribution P over the set S is known, then the expected utility of A_k is

$$E_k = \sum_{i \in L} P(S_i) \, \mu_{ik}.$$

In practical situations the utility function is not given directly. What one has is usually a preference relation which implicitly defines a utility function (using the v. Neumann - Morgenstern theory).

In our framework the situations are the problems of the cases in CB, and the actions are the transformations carried out by T; the probabilities have to be replaced by evidences appropriately (see [4], [9]).

Suppose now that a is an actual problem. Using the same notation as in 3.3., we consider after the observation of some attributes (indexed by J) a minimal focal set $X \subseteq CB$ for which we have the accumulated evidence $\nu_J(X)$. We define for $x \in X \subseteq CB$

$u_{x,T} :=$ utility of applying T to x where $u_{x,T}$ depends on the parameters β_1, β_2 introduced above.

Because all cases of X are indiscernable, it is reasonable to put

$$\mu_J^D(a, x) = \nu_J(x) \cdot \mu_{x,T} \quad \text{for } x \in X.$$

If the configuration task degenerates the classification, we obtain this as a special form of the result from 3.3.

Acknowledgement: The author is indebted to Christoph Globig for helpful discussions.

REFERENCES:

[1]   Dempster, A.: Upper and Lower Probabilities Induced by a Multivalued Mapping, Anuals Math. Stat., 38 (1967), 325-339.

[2]   Kohlas, J.: Modeling Uncertainty with Belief Functions in Numerical Models, Eur. J. of Oper. Res., 40 (1989), 377-388.

[3]   Kohlas, J.: A Mathematical Theory of Hints, Inst. for Aut. and Oper. Res.,Univ. of Fribourg, Tech. Report, 173 (1990).

[4]   Kohlas, J.and Monney, P.-A.: Theory of Evidence - A survey of its Mathematical Foundations, Applications and Computational Aspects, ZOR, 39 (1994), 35 - 68.

[5]   Orponen, P.: Dempster's Rule of Combination is #P-complete, Artif. Intell., 44 (1990), 245 - 253.

[6]   Paulokat, J. and Weß, S.: CABLAN - fallbasierte, nichtlineare, hierarchische Arbeitsplanung, in: Beiträge zum 2. Workshop AK-CBR (Ed. D. Janetzko, Th. Schult), Freiburg 1993.

[7]   Richter, M. M. and Weß, S.: Similarity, Uncertainity and Case-Based Reasoning in PATDEX, in: Authoricated Reasoning. Essays in Honor of Woody Bledsoe (Ed. R. S. Boyer), Kluwer 1991, 249 - 265.

[8]   Shafer, G.: A Mathematical Theory of Evidence, Princeton University Press 1967.

[9]   Strat, T. M.: Decision Analysis Using Belief Functions, Int. J. Approximate Reasoning, 4 (1990), 391 - 418.

[10] Tversky, A.: Features of Similarity, Psych. Review, 84 (1977), 327 - 352.

[11] Weß, S.: PATDEX/2: Ein System zum adaptieren, fallfokussierenden Lernen in technischen Systemen, SEKI - Working Paper SW91/01, Dept. of Comp. Science, Univ. of Kaiserslautern 1991.

[12] Weß, S. and Althoff, K.D. and Richter, M. M.: Topics in Case-Based Reasoning. EWCBR '93, Selected Papers. Springer LNAI 837 (1994).

LINEAR SPACE INDUCTION
IN FIRST ORDER LOGIC WITH RELIEFF

U. Pompe and I. Kononenko
University of Ljubljana, Ljubljana, Slovenia

Abstract

Current ILP algorithms typically use variants and extensions of the greedy search. This prevents them to detect significant relationships between the training objects. Instead of myopic impurity functions, we propose the use of the heuristic based on RELIEF for guidance of ILP algorithms. At each step, in our ILP-R system, this heuristic is used to determine a beam of candidate literals. The beam is then used in an exhaustive search for a potentially good conjunction of literals. From the efficiency point of view we introduce interesting declarative bias which enables us to keep the growth of the training set, when introducing new variables, within linear bounds (linear with respect to the clause length). This bias prohibits cross-referencing of variables in variable dependency tree. The resulting system has been tested on various artificial problems. The advantages and deficiencies of our approach are discussed.

1   Introduction

ILP algorithms typically use variants of the greedy search strategy to overcome the combinatorial explosion during the search for a good hypothesis. The heuristic function that estimates the potential successors of the current state in the search space has a major role in the greedy search. Current ILP algorithms use variants of impurity functions like information gain (Quinlan, 1990; Džeroski, 1991). However, these measures assume that the estimated literals are independent and therefore the greedy search has poor chances of revealing a good hypothesis. To overcome this problem, various extensions of the greedy search are employed: limited backtracking (Quinlan, 1990), beam search (Džeroski, 1991), and stochastic search (Kovačič et al., 1992; Pompe et al., 1993; Kovačič, 1994).

Kira and Rendell (1992a,b) developed an algorithm, called RELIEF, which was shown to be very effective in estimating the quality of attributes in learning of the propositional theories. For example, in the parity problems of various degrees, with a significant number of irrelevant (random) additional attributes, RELIEF is able to correctly estimate the relevance of all attributes in a time proportional to the number of attributes and the square of the number of training instances (this can be further reduced by limiting the number of iterations in RELIEF). While the original RELIEF can deal with discrete and continuous attributes, it can not deal with incomplete data and is limited to two-class problems only. Kononenko (1994) developed an extension of RELIEF, called RELIEFF, that improves the original algorithm by estimating probabilities more reliably and extends it to handle the incomplete and multi-class data sets.

RELIEFF seems to be a promising heuristic function that may overcome the myopia of current inductive learning algorithms, including ILP algorithms. RELIEFF is general, efficient, and reliable enough to guide the search of the learning process. In this paper a system called ILP-R is described. Instead of information gain, ILP-R uses a heuristic function based on RELIEFF for estimating the literals quality at each step during the clause generation. To cope with the combinatorial explosion and with the space requirements in particular, ILP-R uses declarative bias which prohibits cross referencing among variables in the variable dependency tree. This limitation enables the algorithm to keep the growth of the training set within linear bounds. This property is independent of the type of the literal that introduces new variables. It holds even if such literal is nondeterminate. The language bias also limits the hypothesis space and, unfortunately, the algorithm itself becomes incomplete. But still there are many benchmark and real world problems which can be completely and consistently described.

Experiments on a series of artificial data sets are described and the results obtained using RELIEFF based heuristic as a selection criterion are compared to the results obtained by using the information gain.

The paper is organized as follows. In the next section, the original RELIEF is briefly described along with its extended version RELIEFF. Section 2.2 describes the adaptation of RELIEFF for ILP-R. Section 3 describes the adopted limitations of the target hypotheses language. In section 4, we present the implementation of ILP-R. In section 5 we describe the experimental methodology and the results of experiments. In conclusions, the potential breakthroughs are discussed on the basis of the excellent results on artificial data sets. Finally, some future work is proposed.

2 RELIEFF

2.1 RELIEF

The key idea of RELIEF is to estimate attributes according to how well their values distinguish among the instances that are near to each other. For that purpose, given an instance, RELIEF searches for its two nearest neighbors: one from the same class (called *nearest hit*) and the other from a different class (called *nearest miss*). The original algorithm of RELIEF (Kira & Rendell, 1992a,b) randomly selects n training instances, where n is the user-defined parameter, as follows:

```
set all weights W[A] := 0.0;
for i := 1 to n do
    begin
        randomly select an instance R;
        find nearest hit H and nearest miss M;      { topology searching phase}
        for A := 1 to all_attributes do
            W[A] := W[A] - diff(A,R,H)/n + diff(A,R,M)/n:   { updating phase }
    end;
```

where *diff(Attribute,Instance1,Instance2)* calculates the difference between the values of Attribute for two instances. For discrete attributes the difference is either 1 (the values are different) or 0 (the values are equal), while for continuous attributes the difference is the actual difference normalized to the interval $[0, 1]$. Normalization with n guarantees all weights to be in the interval $[-1, 1]$. The weight W[A] is used as an estimate of the quality of attribute A. The function *diff* is used also for calculating the distance between instances to find the nearest neighbors. The total distance is simply the sum of differences over all attributes. In fact, the original RELIEF uses the squared difference, which for discrete attributes (which is the case in ILP, see section 2.2) is equivalent to *diff*.

It is obvious that RELIEF's estimate $W[A]$ of attribute A is an approximation of the following difference of probabilities:

$$W[A] = P(\text{different value of A}|\text{nearest instance from different class})$$

$$-P(\text{different value of A}|\text{nearest instance from same class}) \qquad (1)$$

The rationale of the above formula is that a good attribute should differentiate between instances from different classes and should have the same value for instances from the same class.

Kononenko (1994) has shown that (1) is highly correlated with gini index (Breiman et al., 1984) and that it implicitly uses a kind of normalization for multi valued attributes.

(impurity functions tend to overestimate multi valued attributes and various normalization heuristics are needed to avoid this tendency, see (Quinlan, 1986)). Due to the "nearest instance" condition, we can interpret the RELIEF's estimates as the average over local estimates in the small parts of the instance space. This enables RELIEF to take into account the dependencies between attributes which can be detected in the context of locality. From the global point of view, when using classical impurity functions like information gain, these dependencies are hidden due to the effect of averaging over all training instances, and exactly this makes impurity functions myopic.

Kononenko (1994) developed an extension of RELIEF, called RELIEFF, that improves the original algorithm by estimating probabilities more reliably (by running the outer loop of RELIEFF over all training instances and by using k-nearest hits/misses instead of only one nearest hit/miss) and extends it to deal with incomplete data (by generalizing the function $diff$ to calculate the probability that two instances have different value for a given attribute) and multi-class data sets (by selecting one near miss for each different class instead of one near miss from only one of different classes).

2.2 RELIEFF for ILP-R

RELIEFF as described above is suitable for the propositional representation of training examples. A slightly different approach is needed when estimating the quality of literals which are the candidates for augmenting the current clause under construction. The main difference stems from the fact that while learning in the propositional language we are only interested in the boundaries between different classes. On the other hand, when learning in the first order language we are not searching for boundaries them self but for a theory that explains positive learning instances and does not cover negative ones. A crucial part of RELIEFF is the function that measures the difference (distance) between the training instances.

Given a partial theory with an unfinished clause, one training instance a is represented as an n-ary relation $R(a)$, where n is the number of all variables in the clause. If $n=$arity of the head of the clause, then the relation contains exactly one n-tuple, i.e. the original training instance: $a = \langle a_1, ..., a_n \rangle$. If the body of the clause contains additional variables, the number of n-tuples, corresponding to one original training instance, may be greater than one. If we add to the clause a literal l that does not introduce any new variables, the new relation $R_l(a)$ has the same arity, while the number of n-tuples may decrease. If the new literal l does not "cover" the training instance a, then $|R_l(a)| = 0$. On the other hand, if the literal introduces new variables, the arity of $R_l(a)$ increases to $m = n + r, r > 0$. The number of the corresponding m-tuples may increase or decrease, or remains unchanged.

When evaluating the quality of the new candidate literals a training example a is described with:

- $C_l(a)$ represents the truth value for literal l; literals are treated by RELIEFF as the binary attributes with a corresponding truth value for each training instance. If $C_l(a)$ is true then the new partial clause obtained by adding the literal l to the body is still covering instance a.

- the relation $R_l(a)$ for each new "open" literal l that introduces new variables (negative literals that introduce new variables do not have this information added since they are not in fact introducing any new bindings. Therefore, they are treated as "closed" literals taking into account only their coverage vector C_l).

The difference between two training instances a and b, used by RELIEFF to estimate each candidate literal l, is defined as:

$$diff(a,b) = \frac{1}{\#new\ literals} \times$$

$$\left(\overset{\#closed\ lit.}{\underset{l}{\sum}} difftruth(C_l(a), C_l(b)) + \overset{\#open\ lit.}{\underset{l}{\sum}} diffopen(a,b) \right) \qquad (2)$$

where $difftruth(P1, P2)$ is the difference between the truth values of the two conditions. Diffopen is computed according to the formula (3).

$$diffopen(a,b) = \begin{cases} difftruth(C_l(a), C_l(b)) & ; \quad C_l(a) \neq true \lor C_l(b) \neq true \\ diffattr(R_l(a), R_l(b), m, r) & ; \quad \text{otherwise} \end{cases}$$

$$(3)$$

$diffattr(A, B, m, r)$ is the difference between sets A and B where $r = m - n$ denotes the number of variables introduced by the literal (n is the number of old variables and m represents all variables):

$$diffattr(A, B, m, r) = \frac{1}{m+r} \left[\sum_{i=1}^{m} diffset(A_i, B_i) + \sum_{j=n+1}^{n+r=m} diffset(Cov_j(A), Cov_j(B)) \right]$$

$$(4)$$

A_i (and similarly B_i) represents the set of all bindings of j-th variable (5).

$$A_i = \{x_i | \exists \langle x_1, \ldots, x_i, \ldots, x_m \rangle \in A\} \qquad (5)$$

$Cov_j(X)$, on the other hand, conveys entirely different information. Namely, it represents the provability of variable X_j with respect to the attributional part of the background knowledge (6).

$$Cov_j(X) = \{a_k | \exists \langle x_1, \ldots, x_j, \ldots, x_m \rangle \in X \land p_k(x_j) = true\} \qquad (6)$$

Example 1:

Suppose we are given a background knowledge consisting of predicates p1,p2,p3 and target predicate p.

```
p1(a,b).
p1(a,c).
p1(b,b).
p1(c,a).
p1(c,c).
p2(a).
p2(c).
p3(b).

p(a,b).
p(b,a).
p(b,c).
p(c,b).
```

For partial clause

```
p(X,Y) :-
```

and literal $l = p(X, Z)$ that we are trying to asses the following holds:

$$
\begin{aligned}
A_X(R_l(a,b)) &= \{a\} \\
A_Y(R_l(a,b)) &= \{b\} \\
A_Z(R_l(a,b)) &= \{b,c\} \\
Cov_Z(R_l(a,b)) &= \{att_1, att_2\}
\end{aligned}
$$

where att_1 corresponds to attribute p2 and att_2 to attribute p3, respectively.

When comparing two training instances (open literals only), we first compute the difference in bindings of respective variables. Then, additionally, the difference in attributional coverage of new variable bindings is determined. That is, all new bindings are checked against the 1-ary predicates from the background knowledge for their truth value. This look ahead over attributional background knowledge seems very costly but is in fact quite simple since the coverage of the individual bindings can be computed in advance. After all the information about the bindings and coverage has been collected $diffset$ only computes the difference of two sets (7). Let $\mathcal{U}$ be the universal set (all possible elements) of A_x and B_x: $A_x \subseteq \mathcal{U}, B_x \subseteq \mathcal{U}$. Then:

$$
diffset(A_x, B_x) = \frac{1}{|\mathcal{U}|} \times \sum_{u_i \in \mathcal{U}} difftruth(u_i \in A_x, u_i \in B_x) \tag{7}
$$

RELIEFF in ILP-R uses the equation (2) when searching for the nearest hits/misses, and functions $difftruth$ and $diffopen$ when updating the estimates of the quality of the closed and open candidate literals, respectively.

As the equations show all functions rely on the definition of $difftruth$. To satisfy our interest in covering theories as described at the beginning of this section, ILP-R uses different $difftruth$ functions when in topology searching phase of RELIEF and when it updates the estimates of quality (see the algorithm in section 2.1). Table 1 shows the different definitions. Note that because of the asymmetry the order of arguments is important.

$difftruth(X,Y)$			
X	Y	neighbourhood	quality
0	0	0	0
0	1	1	0
1	0	1	1
1	1	0	0

Table 1: Definition of $difftruth$ in topology searching phase and updating phase.

3 Limitations of the hypothesis language

When we are introducing new variables in the clause we are often confronted with combinatorial explosion. All new bindings must be remembered and when the "open" literals are not determinate this results in an exponential growth of the training set. To cope with this problem we are introducing a declarative bias which is used by ILP-R to keep the growth of the training set within linear bounds with respect to the clause length.

First we need some basic notions. Let $\mathcal{B}$ represent the background knowledge; the set of all predicates from which we are trying to built our hypothesis (we are considering only function-free horn clause logic).

Definition 1 *Open literal L is the positive literal in the body of clause C for which holds: there exists such an ordering of literals in the body that the following is true:*

1. *L has exactly one old variable; a variable that was introduced by some other open literal before L or it belongs to the head.*

2. *L introduces at least one new variable*

Such literal, when added to the body, expands the variable space as well as the literal space, since many new logic expressions are possible on larger variable space. The head is also an open literal. We are only considering clauses where new variables are introduced exclusively by open literal. With $Arg(L)$ we denote the set of variables used as arguments of predicate $p \in \mathcal{B}$; p is primary functor in L.

Definition 2 *Variable X depends on variable Y ($X \neq Y$) if one of the following holds:*

- *There exists such an open literal L that both X and Y belong to it's argument set ($X \in Arg(L) \wedge Y \in Arg(L)$), and Y is the old variable of literal L. We say X it "the neighbour variable" of Y.*

- *If there exists some Z such that X depends on Z and Z depends on Y or there exists such open literal L that holds $Z \in Arg(L) \wedge Y \in Arg(L)$ and X depends on Z then X depends on Y.*

With the above definition we can now, for every clause, build a variable dependency graph where nodes are sets of variables introduced by some open literal in the clause and arcs represent the neighbourhood between them. Note, however, that the definition of dependency implies that the variable dependency graph is always a tree. It can be observed that closed literals do not affect the variable dependency tree.

Definition 3 *Literal L is i-dependable with respect to a given clause if there exist a path of length i in the variable dependency tree that connects all argument variables.*

Now we can finally define our declarative bias. As our hypothesis description language we use:

Definition 4 (d-dependable language)

   $\mathcal{L}_d = \{ \mathcal{C} \; ; \mathcal{C}$ *is function free horn clause* $\wedge \forall L_i \in \mathcal{C} \Rightarrow L_i$ *is at most d-dependable$\}$*

Unfortunately, this hierarchy of languages is far from complete, even when d goes to infinity. Still, many of the benchmark problems, known from the literature, as well as many real world problems can be described by using one of these languages, even for a small d. In ILP-R we use $\mathcal{L}_2$ as our description language.

If the completion is what we lost then what do we gain. It can be easily shown that d-dependable languages have some interesting properties.

Theorem 1 (Space bound) *For every finite d there exists such K that it holds:*

$$\forall \mathcal{C}_1, \mathcal{C}_2 \in \mathcal{L}_d \wedge \mathcal{C}_1 \subset \mathcal{C}_2 \Rightarrow Space(\mathcal{C}_2) \leq Space(\mathcal{C}_1) + K * (|\mathcal{C}_2| - |\mathcal{C}_1|) * M$$

where $Space(C_i)$ denotes the size of the training set representing all possible binding of variables and M denotes the maximal size of any predicate p from $\mathcal{B}$.

Proof:

We are only going to sketch the proof. Closed literals obviously do not impose any spatial requirements since they only cancel existent bindings in the current training set. Our concern are open literal which introduce new bindings. It can be observed that the introduction of new variables is completely independent step for all variables that are more than d steps away in a variable dependency tree (on a descending path). We can, therefore, coalesce a tree and make it a mesh. Only the binding-validity information should be remembered and its size is bounded by constant d. From everything said above we can conclude that the use of $\mathcal{L}_d$ enables us to keep the growth of training set within linear boundaries. ∎

4 ILP-R

Finally, we can describe our ILP-R system. The background knowledge as well as the target predicate are entered as a set of ground facts. The system is using covering approach, when building a theory, similar to that of FOIL (Quinlan, 1990), mFOIL (Džeroski, 1991), SFOIL (Pompe, 1993), and many other top down learners.

At the outer level we search for an almost consistent clause (when learning in the presence of noise, the consistency criteria must be slightly loosened), and then remove all positive training instances covered by that clause from the training set.

At the inner level, when building a clause, we are using RELIEFF based heuristic to asses the quality of literals. The most promising literals are then gathered in a beam which is then exhaustively searched for the most informative phrase (conjunction of literals). Only the least restrictive (one covering the most positive examples) literal is then added to the clause and the cycle is repeated as long as the partial clause is not consistent or some predefined upper bound is reached.

When randomly selecting training instances for our RELIEFF based estimator only positive ones are taken into account. This and the definition of the $difftruth$ in fact changes the probability that system is trying to approximate. Instead of (1) we have now the following formula:

$$W[L] = P(\text{L covers positive and not negative instance}|\text{nearest negative example})$$

$$-P(\text{different coverage of L}|\text{nearest positive example}) \qquad (8)$$

This estimate assesses more properly the quality of the literal than the original definition for the reasons mentioned in section 2.2.

Some more sophisticated methods for dealing with noise have yet to be implemented into the system; and this weakness is exposed very quickly when learning from noisy data.

5   Experimental results

In this section we give results of experiments in several attributional and relational artificial data sets and one real-world data set.

5.1   Experimental methodology

When not specified otherwise, the split between the training and the testing data was left as it was, so that from that point of view our tests are compatible with those from the literature. All attributional problems were transformed uniformly into relational ones by the following method. The training instances were renamed into the target predicate of the corresponding arity. Domains of individual attributes were created and attributes were added to the background knowledge in the form of unary predicates collecting their v values into v 1-ary relations.

Example 2:

The instance "color = blue & shape = triangle & class =invalid" can be described with the clause:

```
class(Color,Shape,Class) :-
        blue(Color),
        triangle(Shape),
        invalid(Class)
```

As a matter of fact, the same effect can be achieved with the literals of the type "$X = const.$" where X is a variable and *const.* is a constant from domain $Dom(X)$.

When learning from such domains we prohibited the use of recursion, partly because we wanted to test how our algorithm performs in the situation when dealing with similarly constraint language, but mostly because it would not be feasible to allow recursion since target predicates had quite a large arity.

Table 2: Basic description of attributional artificial data sets

domain	#class	#atts.	#val/att.	# instances	maj.class (%)	entropy(bit)
PAR2	2	12	2.0	200	54	0.99
PAR3	2	13	2.0	200	54	0.99
PAR4	2	14	2.0	400	50	1.00

Besides the classification accuracy, we measured also the average information score (Kononenko & Bratko, 1991). This measure eliminates the influence of prior probabilities and appropriately treats probabilistic answers of a classifier. The average information score is defined as:

$$Inf = \frac{\sum_{i=1}^{\#testing\ instances} Inf_i}{\#testing\ instances} \tag{9}$$

where the information score of the classification of i-th testing instance is defined by:

$$Inf_i = \begin{cases} -\log_2 P(Cl_i) + \log_2 P'(Cl_i), & P'(Cl_i) \geq P(Cl_i) \\ -(-\log_2(1 - P(Cl_i)) + \log_2(1 - P'(Cl_i))), & P'(Cl_i) < P(Cl_i) \end{cases}$$

Cl_i is the class of the i-th testing instance, $P(Cl)$ is the prior probability of class Cl and $P'(Cl)$ the probability returned by a classifier.

5.2 Attributional artificial data sets

We generated three artificial data sets where the background knowledge consisted of attributes (predicates with arity = 1) only:

PAR2: Parity problem with two significant binary attributes and 10 random binary attributes. 5 % of randomly selected instances were labelled with the wrong class. This problem is hard as there is a lot of attributes with equal score when evaluated with the myopic evaluation function, such as information gain.

PAR3: Same as PAR2 except that there were three significant attributes for the parity relation which makes the problem harder.

PAR4: Same as PAR2 except that there were four significant attributes for the parity relation which makes the problem the hardest among the parity problems used in our experiments.

Table 3: Classification accuracy of the learning systems on artificial data sets

domain	ILP-R	ILP-R (Inf. gain)	Assistant	naive Bayes	k-NN
PAR2	92.7±2.1	88.7± 4.42	69.9± 5.7	54.6±4.2	78.3±3.9
PAR3	89.4±6.5	57.4± 9.87	68.1±13.0	55.6±5.7	57.9±3.9
PAR4	93.6±2.2	64.7±11.76	57.9± 6.8	54.8±4.2	62.4±3.1

Table 4: Average information score of the learning systems on artificial data sets

domain	ILP-R	ILP-R (Inf. gain)	Assistant	naive Bayes	k-NN
PAR2	0.84±0.05	0.77±0.09	0.28±0.11	0.07±0.02	0.27±0.03
PAR3	0.79±0.14	0.14±0.20	0.27±0.22	0.03±0.05	0.10±0.04
PAR4	0.87±0.04	0.29±0.24	0.11±0.10	0.06±0.02	0.10±0.03

The basic characteristics of the artificial data sets are listed in table 2. We compared the performance of ILP-R with the naive Bayesian classifier, the K-nearest neighbour algorithm and the inductive learning system Assistant that generates decision trees (Cestnik et al., 1987) which was extended with the m-estimate of probabilities for the decision trees generation and pruning (Cestnik & Bratko, 1991). We also compared the performance of ILP-R (relief) with results obtained when information gain was used as the literal estimator. The results are given in tables 3 (accuracy) and 4 (information score).

In the PARx domains the performance of the naive Bayesian classifier is poor, due to the strong dependencies between attributes. ILP-R, on the other hand performed as expected. It's RELIEFF based estimate is capable of detecting strong dependencies between attributes resulting in correct literals to be passed to the phrase-building phase. The results could have been even better if ILP-R had some basic noise handling technique built in. In fact, the system almost always builds consistent and complete theory which is than unfortunately augmented with clauses covering noisy training examples.

It is interesting to see how well the information gain behaves in the environment with such strong dependencies. In PAR2 domain information gain performed surprisingly well but this is due to the fact that when learning within first order logic xor problem becomes simple. Namely, matching literal $(X = Y)$ is sufficient for describing it. On the other two domains the information gain performed as expected, that is badly, because of it's myopy.

5.3   Relational artificial data sets

The system has been tested on a series of classical ILP benchmark domains known from the literature (Quinlan, 1990). We were trying to induce the definitions of the following predicates: *canreach, member, arch, eastbound, append, reverse, qsort*. ILP-R was able to induce correct hypothesis for the first two predicates. It found very short (one clause) descriptions of the predicates *arch* and *eastbound*. *Append* gave it some trouble (redundant clause was also induced) while *reverse* and *qsort* were too hard. The failures are largely due to the lack of expressiveness of the description language. Especially this is true in the case of *qsort*, since this predicate can not be expressed in an d-dependable language. When using information gain we achieved similar results on this set of benchmark problems.

6   Discussion

RELIEFF is an efficient heuristic estimator of the attribute's quality that is able to deal with data sets with dependent and independent attributes. The extensions implemented in RELIEFF enable it to deal with noisy, incomplete, and multi-class data sets. It seems that RELIEFF based heuristic is capable of detecting literal dependencies and is therefore appropriate for problems where such dependencies exist (i.e. xor). There are two major advantages of the RELIEFF based ILP learner. Because RELIEFF seems to be less often fooled than other myopic heuristics we can rely more strongly on the selected beam and perform exhaustive search without major performance penalty. Second advantage is the use of attributional part of the background knowledge, which seemed to be neglected in current ILP systems. Comparatively, propositional learners usually perform better when learning from the same problem description, therefore, using information hidden in the attributes to a greater extent. By using RELIEFF in ILP, we are hoping to change that.

Acknowledgements

The extensions and experiments with Assistant were performed by Edi Šimec. This work was supported by the Slovenian Ministry of Science and Technology.

References

1. Breiman L., Friedman J.H., Olshen R.A. & Stone C.J. (1984) *Classification and Regression Trees*, Wadsworth International Group.

2. Cestnik B. & Bratko I. (1991) On estimating probabilities in tree pruning, *Proc. European Working Session on Learning*, (Porto. March 1991), Y.Kodratoff (ed.), Springer Verlag. pp.138-150.

3. Cestnik, B., Kononenko, I. & Bratko, I. (1987) ASSISTANT 86 : A knowledge elicitation tool for sophisticated users. In: I. Bratko & N. Lavrač (eds.), *Progress in Machine Learning*. Wilmslow, England: Sigma Press.

4. Džeroski S. (1991) Handling noise in inductive logic programming, M.SC. Thesis, University of Ljubljana, Faculty of electrical engineering & computer science, Ljubljana, Slovenia.

5. Kira K. & Rendell L. (1992a) A practical approach to feature selection, *Proc. Intern. Conf. on Machine Learning* (Aberdeen, July 1992) D.Sleeman & P.Edwards (eds.), Morgan Kaufmann, pp.249-256.

6. Kira K. & Rendell L. (1992b) The feature selection problem: traditional methods and new algorithm. *Proc. AAAI'92*, San Jose, CA. July 1992.

7. Kononenko I. (1994) Estimating attributes: Analysis and extensions of RELIEF. *Proc. European Conf. on Machine Learning* (Catania, April 1994), L. De Raedt & F.Bergadano (eds.), Springer Verlag. (in press)

8. Kononenko, I. & Bratko, I. (1991) Information based evaluation criterion for classifier's performance. *Machine Learning*, 6:67-80.

9. Kovačič M. (1994) Stochastic inductive logic programming, Ph.D. Thesis, University of Ljubljana, Faculty of electr. eng. & computer sc., Ljubljana, Slovenia.

10. Kovačič, M., Lavrač, N., Grobelnik, M., Zupanič, D., Mladenič, D. (1992) Stochastic Search in Inductive Logic Programming. *Proceedings of the tenth European Conference on Artificial Intelligence*, Vienna, 1992.

11. Mladenič D. (1993) Combinatorial optimization in inductive concept learning. *Proc. 10th Intern. Conf. on Machine Learning*. (Amherst, June 1993), Morgan Kaufmann. pp. 205-211

12. Muggleton S. (ed.) (1992) *Inductive Logic Programming*. Academic Press.

13. U.Pompe, M.Kovačič & I.Kononenko (1993) SFOIL: Stochastic approach to inductive logic programming. *Proc. Slovenian Conf. on Electrical Engineering and Computer Science*. Portorož, Slovenia, Sept. 1993, pp 189-192.

14. Quinlan R. (1986) Induction of decision trees, *Machine Learning*. 1:81-106.

15. Quinlan R. (1990) Learning logical definitions from relations, *Machine Learning*. 5:239-266.

16. Smyth P., Goodman R.M. & Higgins C. (1990) A hybrid Rule-based Bayesian Classifier, *Proc.European Conf. on Artificial Intelligence*, Stockholm. August, 1990. pp. 610-615.

INDUCTION OF DECISION TREES USING RELIEFF

I. Kononenko and E. Simec
University of Ljubljana, Ljubljana, Slovenia

Abstract

In the context of machine learning from examples this paper deals with the problem of estimating the quality of attributes with and without dependencies between them. Greedy search prevents current inductive machine learning algorithms to detect significant dependencies between the attributes. Recently, Kira and Rendell developed the RELIEF algorithm for estimating the quality of attributes that is able to detect dependencies between attributes. We show strong relation between RELIEF's estimates and impurity functions, that are usually used for heuristic guidance of inductive learning algorithms. We propose to use RELIEFF, an extended version of RELIEF, instead of myopic impurity functions. We have reimplemented Assistant, a system for top down induction of decision trees, using RELIEFF as an estimator of attributes at each selection step. The algorithm is tested on several artificial and several real world problems. Results show the advantage of the presented approach to inductive learning and open a wide range of possibilities for using RELIEFF.

1 Introduction

Machine learning encompasses algorithms that learn a task from a series of examples. There are various subfields of machine learning, such as inductive learning from examples, reinforcement learning, case-based learning, explanation-based learning, inductive logic programming, computational learning theory, and automated discovery. It is arguable whether learning of neural networks and statistical classification tehniques can be put under the Machine Learning umbrella. For a collection of relevant papers that illustrate the development of the field see (Dietterich & Shavlik, 1990). An extensive bibliography of the field can be found in (Michalski & Tecuci, 1994). In this contribution we are concerend with inductive learning from examples.

Typical inductive learning algorithm uses a greedy search strategy to overcome the combinatorial explosion during the search for good hypotheses. The heuristic functions that estimate the potential successors of the current state in the search space play a major role in the greedy search. Current inductive learning algorithms use variants of impurity functions like information gain, gain ratio (Quinlan, 1986), gini-index (Breiman et al., 1984), distance measure (Mantaras, 1989), weight of evidence (Michie & Al Attar, 1992), and j-measure (Smyth & Goodman, 1990). However, all these measures assume that attributes are independent and therefore in domains with strong dependencies between attributes the greedy search has poor chances of revealing a good hypothesis.

Recently, Kira and Rendell (1992a,b) developed an algorithm called RELIEF, which was shown to be very efficient in estimating the quality of attributes. For example, in the parity problems of various degrees with a significant number of irrelevant (random) additional attributes RELIEF is able to correctly estimate the relevance of all attributes in a time proportional to the number of attributes and the square of the number of training instances. While the original RELIEF can deal with discrete and continuous attributes, it can not deal with incomplete data and is limited to two-class problems only. Kononenko (1994) developed an extension of RELIEF called RELIEFF that improves the original algorithm. RELIEFF estimates probabilities more reliably and efficiently handles incomplete and multi-class data sets while the complexity of the algorithm remains the same.

Kira and Rendell used RELIEF as a preprocessor to eliminate irrelevant attributes from data description before learning. RELIEFF seems to be a promising heuristic function that may overcome the myopia of current inductive learning algorithms. It is general, efficient, and reliable enough to guide the search in the learning process. In this paper a reimplementation of Assistant learning algorithm for top down induction of decision trees (Cestnik et al., 1987) is described, named Assistant-R. Instead of information gain, Assistant-R uses RELIEFF as a heuristic function for estimating the attributes quality at each step during the tree generation. Experiments on a series of artificial and real-world data sets are described and the results obtained using RELIEFF as a selection criterion are compared to results obtained by using information gain instead of RELIEFF and to the results of the naive Bayesian classifier that uses the m-estimate of probabilities (Cestnik, 1990).

In addition we discuss also two other approaches:

- the use of RELIEFF as a preprocessor for eliminating irrelevant attributes, as suggested by Kira and Rendell (1992a,b);

- dealing with multiple-class problems by generating a series of decision trees, each for one class; this is one possible approach to the extension of RELIEF to deal with decision problems with more than two classes, as suggested by Kira and Rendell (1992a,b).

In the next section we describe the original RELIEF along with its interpretation and its extended version RELIEFF. In section 3, the reimplementation of Assistant called Assistant-R is described. Section 4 describes experiments and compares the results of the different algorithms. We also briefly discuss two approaches mentioned above that were suggested by Kira and Rendell. In conclusions, the potential breakthroughs are discussed on the basis of the excellent results on artificial data sets.

2 Algorithm RELIEFF

2.1 Overview of RELIEF

Recently, Kira & Rendell (1992a,b) developed an algorithm that is able to efficiently estimate the quality of attributes with strong interdependencies that can be found for example in parity problems. The key idea of RELIEF is to estimate attributes according to how well their values distinguish among the instances that are near to each other. For that purpose, given an instance, RELIEF searches for its two nearest neighbors: one from the same class (called *nearest hit*) and the other from a different class (called *nearest miss*). The original algorithm of RELIEF randomly selects n training instances, where n is the user-defined parameter, as follows:

set all weights $W[A] := 0.0$;
for i := 1 **to** n **do**
   begin
      randomly select an instance R;
      find nearest hit H and nearest miss M;
      for A := 1 **to** #all_attributes **do**
         $W[A] := W[A] - \text{diff}(A,R,H)/n + \text{diff}(A,R,M)/n$;
   end;

where *diff(Attribute,Instance1,Instance2)* calculates the difference between the values of Attribute for two instances. For discrete attributes the difference is either 1 (the values are different) or 0 (the values are equal), while for continuous attributes the difference is the actual difference normalized to the interval $[0,1]$. Normalization with n guarantees all weights $W[A]$ to be in the interval $[-1,1]$.

The weights are estimates of the quality of attributes. The rationale of the formula for updating the weights is that a good attribute should have the same value for instances from the same class (subtracting the difference $diff(A, R, H)$) and should differentiate between instances from different classes (adding the difference $diff(A, R, M)$).

The function *diff* is used also for calculating the distance between instances to find the nearest neighbors. The total distance is simply the sum of differences over all attributes (in fact original RELIEF uses the squared difference, which for discrete attributes is equivalent to *diff*. In all our experiments, there was no significant difference between results using *diff* or squared difference). If N is the number of all training instances then the complexity of the above algorithm is $O(n \times N \times \#all\_attributes)$.

2.2   The relation with impurity functions

In the following derivation we show that RELIEF's estimates are strongly related to impurity functions. It is obvious that RELIEF's estimate $W[A]$ of attribute A is an approximation of the following difference of probabilities:

$$W[A] = P(\text{different value of A}|\text{nearest instance from different class})$$
$$- P(\text{different value of A}|\text{nearest instance from same class}) \qquad (1)$$

If we eliminate from (1) the requirement that the selected instance is *the nearest*, the formula becomes:

$$W[A] = P(\text{different value of A}|\text{different class})$$
$$- P(\text{different value of A}|\text{same class}) \qquad (2)$$

If we rewrite

$$P_{eqval} = P(\text{equal value of A})$$
$$P_{samecl} = P(\text{same class})$$

and

$$P_{samecl|eqval} = P(\text{same class}|\text{equal value of A})$$

we obtain using Bayes rule:

$$W[A] = \frac{P_{samecl|eqval}P_{eqval}}{P_{samecl}} - \frac{(1 - P_{samecl|eqval})P_{eqval}}{1 - P_{samecl}}$$

In the following we will need two equalities:

$$P_{samecl} = \sum_{C} P(C)^2$$

$$P_{samecl|equal} = \sum_V \left(\frac{P(V)^2}{\sum_V P(V)^2} \times \sum_C P(C|V)^2 \right)$$

The former equality is trivial. The derivation of the latter equality follows:

$$
\begin{aligned}
P_{samecl|equal} &= P(C_1 = C_2 | V_1 = V_2) \\
&= \frac{P(C_1 = C_2 \ \& \ V_1 = V_2)}{P(V_1 = V_2)} \\
&= \frac{\sum_V P(C_1 = C_2 \ \& \ V_1 = V_2 = V)}{\sum_V P(V)^2} \\
&= \frac{\sum_V \left(P(V_1 = V_2 = V) P(C_1 = C_2 | V_1 = V_2 = V) \right)}{\sum_V P(V)^2} \\
&= \frac{\sum_V \left(P(V)^2 \sum_C P(C_1 = C_2 = C | V_1 = V_2 = V) \right)}{\sum_V P(V)^2} \\
&= \frac{\sum_V \left(P(V)^2 \sum_C P(C|V)^2 \right)}{\sum_V P(V)^2} \\
&= \sum_V \left(\frac{P(V)^2}{\sum_V P(V)^2} \times \sum_C P(C|V)^2 \right)
\end{aligned}
$$

$$(3)$$

Using the above equalities we obtain:

$$
\begin{aligned}
W[A] &= \frac{P_{equal} \times Gini'(A)}{P_{samecl}(1 - P_{samecl})} \\
&= constant \times \sum_V P(V)^2 \times Gini'(A)
\end{aligned}
\qquad (4)
$$

where

$$Gini'(A) = \sum_V \left(\frac{P(V)^2}{\sum_V P(V)^2} \times \sum_C P(C|V)^2 \right) - \sum_C P(C)^2 \qquad (5)$$

is highly correlated with the gini-index (Breiman et al., 1984) for classes C and values V of attribute A. The only difference is that instead of the factor

$$\frac{P(V)^2}{\sum_V P(V)^2}$$

the gini-index uses

$$\frac{P(V)}{\sum_V P(V)} = P(V)$$

The probability $\sum_V P(V)^2$ that two instances have the same value of attribute A in eq. (3) is a kind of normalization factor for multi valued attributes. Impurity functions tend to overestimate multi valued attributes and various normalization heuristics are needed to avoid this tendency (e.g. gain ratio (Quinlan, 1986) and binarization of attributes (Cestnik et al., 1987)). Equation (3) shows that RELIEF implicitly uses such a normalization.

In the above derivation we eliminated the *"nearest instance"* condition from the probabilities. If we put it back we can interpret RELIEF's estimates as the average over local estimates in smaller parts of the instance space. This enables RELIEF to take into account dependencies between attributes which can be detected in the context of locality. From the global point of view, these dependencies are hidden due to the effect of averaging over all training instances, and exactly this makes impurity functions myopic.

2.3   Extensions of RELIEF

RELIEF can deal with discrete and continuous attributes. However, it can not deal with incomplete data and is limited to two-class problems only. Kononenko (1994) developed an extension of RELIEF, called RELIEFF, that improves the original algorithm by estimating probabilities more reliably and extends it to deal with incomplete and multi-class data sets. A brief description of the extensions follows.

2.3.1   Reliable probability approximation

Parameter n in the algorithm RELIEF, described in section 2.1, represents the number of instances for approximating probabilities in eq. (1). The larger n implies more reliable approximation. The obvious choice, adopted in RELIEFF for relatively small number of training instances (up to one thousand), is to run the outer loop of RELIEF over all available training instances.

The selection of the nearest neighbors is of crucial importance in RELIEF. The purpose is to find the nearest neighbors with respect to important attributes. Redundant and noisy attributes may strongly affect the selection of the nearest neighbors and therefore the estimation of probabilities with noisy data becomes unreliable. To increase the reliability of the probability approximation RELIEFF searches for k nearest hits/misses instead of only one near hit/miss and averages the contribution of all k nearest hits/misses. It was shown that this extension significantly improves the reliability of estimates of attributes' qualities (Kononenko, 1994). To overcome the problem of parameter tuning, in all our experiments k was set to 10 which, empirically, gives satisfactory results. In some problems significantly better results can be obtained with tuning (as is typical for the majority of machine learning algorithms). We will mention some of results of this kind in section 5.

2.3.2 Incomplete data

To enable RELIEF to deal with incomplete data sets, the function *diff(Attribute,Instance1, Instance2)* in RELIEFF is extended to missing values of attributes by calculating the probability that two given instances have different values for the given attribute:

- if one instance (e.g. I1) has unknown value:

$$diff(A, I1, I2) = 1 - P(value(A, I2)|class(I1))$$

- if both instances have unknown value:

$$diff(A, I1, I2) = 1 - \sum_{V}^{\#values(A)} \left(P(V|class(I1)) \times P(V|class(I2)) \right)$$

The conditional probabilities are approximated with relative frequencies from the training set.

2.3.3 Multi-class problems

Kira and Rendell (1992a,b) claim that RELIEF can be used to estimate the attributes' qualities in data sets with more than two classes by splitting the problem into a series of 2-class problems. This solution seems unsatisfactory. To use it in practice, RELIEF should be able to deal with multi class problems without any prior changes in the knowledge representation that could affect the final outcomes.

Instead of finding one near miss M from a different class. RELIEFF finds one near miss $M(C)$ for each different class C and averages their contribution for updating the estimate $W[A]$. The average is weighted with the prior probability of each class:

$$W[A] := W[A] - diff(A, R, H)/n + \sum_{C \neq class(R)} [P(C) \times diff(A, R, M(C))] / n$$

The idea is that the algorithm should estimate the ability of attributes to separate each pair of classes regardless of which two classes are closest to each other. Note that the complexity of RELIEFF is $O(N^2 \times \#all\_attributes)$, where N is the number of training instances.

3 Induction of decision trees with Assistant-R

Assistant-R is a reimplementation of the Assistant learning system for top down induction of decision trees (Cestnik et al., 1987). The basic algorithm goes back to CLS
(Concept Learning System) developed by Hunt et al. (1966) and reimplemented by
several authors (see (Quinlan, 1986) for an overview). In the following we describe the
main features of Assistant and the differences implemented in Assistant-R.

3.1 Binarization of attributes

The algorithm generates binary decision trees. At each decision step the binarized version of each attribute is selected that maximizes the information gain of the attribute.
For continuous attributes a decision point is selected that maximizes the attribute's
information gain. For discrete attributes a heuristic greedy algorithm is used to find
the locally best split of attribute's values into two subsets.

The purpose of the binarization is to reduce the replication problem and to strengthen
the statistical support for generated rules. Many top down induction of decision tree
algorithms construct binary decision trees in order to prevent too detailed splits in
one decision node (e.g. Hunt et al., 1966; Breiman et al., 1984; Cestnik et al., 1987;
Quinlan, 1986; Fayyad, 1991; Fayyad & Irani, 1992).

3.2 Decision tree pruning

Prepruning and postpruning techniques are used for pruning off unreliable parts of
decision trees. For prepruning, three user-defined thresholds are provided: minimal
number of training instances, minimal attributes information gain and maximal probability of majority class in the current node. For postpruning, the method developed by
Niblett and Bratko (1986) is used that uses Laplace's law of succession for estimating
the expected classification error of the current node commited by pruning/not pruning
its subtree.

3.3 Incomplete data handling

During learning, training instances with a missing value of the selected attribute are
weighted with probabilities of each attribute's value conditioned with a class label.

During classification, instances with missing values are weighted with unconditional probabilities of attribute's values.

3.4   Naive Bayesian classifier

For each internal node in a decision tree eventually a third successor appears labelled with attribute's values for which no training instances are available. For such "null leaves", the naive Bayesian formula is used to calculate the probability distribution in the leaf by using only attributes that appear in the path from the root to the leaf:

$$P(C|A_{root}..A_{leaf}) = P(C) \prod_A \frac{P(C|A)}{P(C')} \tag{6}$$

Note that this calculation is done off-line, i.e. during the learning phase. For classification, the "null" leaves are already labeled with the calculated class probability distribution and are used for classification in the same manner as ordinary leaves.

3.5   The differences implemented in Assistant-R

The main difference between Assistant and its reimplementation Assistant-R is that RELIEFF is used for attribute selection. In addition, wherever appropriate, instead of the relative frequency, Assistant-R uses the m-estimate of probabilities, which was shown to often significantly increase the performance of machine learning algorithms (Cestnik, 1990; Cestnik & Bratko, 1991). For prior probabilities Laplace's law of succession is used:

$$P_a(X) = \frac{N(X) + 1}{N + \#\_of\_possible\_outcomes} \tag{7}$$

where N is the number of all trials and $N(X)$ the number of trials with the outcome X. These prior probabilities are then used in the m-estimate of conditional probabilities:

$$P(X|Y) = \frac{N(X\&Y) + m \times P_a(X)}{N(Y) + m} = \frac{N(X\&Y)}{N(Y) + m} + \frac{m \times P_a(X)}{N(Y) + m} \tag{8}$$

The parameter m trades off between the contributions of the relative frequency and the prior probability.

In our experiments, the parameter m was set to 2 (this setting is usually used as default and, empirically, gives satisfactory results (Cestnik, 1990; Cestnik and Bratko, 1991) although with tuning in some problem domains better results may be expected). The m-estimate is used in the naive Bayesian formula (5), for postpruning instead of Laplace's law of succession as proposed by Cestnik and Bratko (1991), and for RELIEFF's estimates of probabilities. In eq. (1) we can use probabilities from the root of the tree as an estimate of prior probabilities for a lower internal node t with $n(t)$ corresponding training instances:

$$
\begin{aligned}
W[A] = & \left(\frac{n(t)}{n(t)+m} \times P(\text{different value of A}|\text{nearest miss, } t) + \right. \\
& \left. \frac{m}{n(t)+m} \times P(\text{different value of A}|\text{nearest miss}) \right) - \\
& \left(\frac{n(t)}{n(t)+m} \times P(\text{different value of A}|\text{nearest hit, } t) + \right. \\
& \left. \frac{m}{n(t)+m} \times P(\text{different value of A}|\text{nearest hit}) \right)
\end{aligned}
\tag{9}
$$

4 Experiments on various data sets

In this section we describe experimental methodology and results on a series of artificial data sets and a series of real world data sets.

4.1 Experimental methodology

We performed a series of experiments with Assistant-R and compared its performance to the following algorithms:

Assistant-I: A variant of Assistant-R that instead of RELIEFF uses information gain for the selection criterion, as does Assistant. However, the other differences to Assistant remain (m-estimate of probabilities). This algorithm enables us to evaluate the contribution of RELIEFF. The parameters for Assistant-I and Assistant-R were fixed throughout the experiments (no prepruning, postpruning with m = 2).

Naive Bayesian Classifier: A classifier that uses the naive Bayesian formula (5) to calculate the probability of each class given the values of all attributes and assuming the conditional independence of the attributes. A new instance is classified into the class with maximal calculated probability. The m-estimate of probabilities was used and the parameter m was set to 2 in all experiments.

In addition we tested also two additional versions of Assistant that implement ideas proposed by Kira and Rendell (1992a,b) (for brevity, we include here only a qualitative evaluation of the results of these two approaches and ommit them from tables):

Elimination of attributes using RELIEFF: A variant of Assistant-I. that uses RELIEFF as a preprocessor to evaluate all attributes and eliminates from learning the attributes whose score is less than a user defined parameter τ.

It turned out that the results are very sensitive to the setting of the threshold τ. Kira and Rendell proposed the default value 0.1 which typically works on artificial data sets, while in real world problems the optimal τ is between 0.0 and 0.1. If a non-optimal τ is selected the results of this algorithm may be much worse than the results of Assistant-R. With the optimal τ this algorithm achieves the performance of Assistant-R. Note that τ must be selected while considering the training set only, which seems to be a non-trivial task. Cross validation on training set for parameter tuning seems to be appropriate at the cost of the increased complexity.

Assistant-R2: A variant of Assistant-R that generates one decision tree for each class in a problem domain, considering the union of the remaining classes as one class. This variant of Assistant-R does not require the extension of RELIEF to multi-class problems. The classification of an unseen case is performed with all decision trees. However, in conflict situations (if no tree or more than one tree "fires") the normal decision tree of Assistant-R can be used for classification.

In all our experiments there was no significant difference between the performance of Assistant-R and Assistant-R2. Assistant R is much more efficient as Assistant-R2 has to generate a series of decision trees. On the other hand. if there is no conflict (and in the majority of cases there was not) during the classification of an unseen case, the classification is more reliable and a series of decision trees may provide different points of view to explain the classification.

Each experiment on each data set was performed 10 times by randomly selecting 70% of instances for learning and 30% for testing and the results were averaged. Each system used the same subsets of instances for learning and for testing in order to provide the same experimental conditions. The only exception from the above methodology were the experiments in the finite element mesh design problem. where the experimental methodology was dictated by previous published results. as described in subsection 5.4.

Besides the classification accuracy, we measured also the average information score (Kononenko & Bratko, 1991). This measure eliminates the influence of prior probabilities and appropriately treats probabilistic answers of a classifier. The average information score is defined as:

$$Inf = \frac{\sum_{i=1}^{\#testing\ instances} Inf_i}{\#testing\ instances} \tag{10}$$

where the information score of the classification of i-th testing instance is defined by:

$$
Inf_i = \begin{cases} -\log_2 P(Cl_i) + \log_2 P'(Cl_i), & P'(Cl_i) \geq P(Cl_i) \\ -(-\log_2(1 - P(Cl_i)) + \log_2(1 - P'(Cl_i))), & P'(Cl_i) < P(Cl_i) \end{cases}
$$

Cl_i is the class of the i-th testing instance, $P(Cl)$ is the prior probability of class Cl and $P'(Cl)$ the probability returned by a classifier.

4.2   Artificial data sets

We generated several data sets in order to compare the performance of various algorithms:

INF1: Domain with three independent informative binary attributes for each of the three classes and with three random binary attributes. The learner should detect which three attributes are informative which is a relatively easy task.

INF2: Domain obtained from INF1 by replacing each informative attribute with two attributes whose values define the value of the original attribute with XOR relation. For this problem, the learner should detect six important attributes and the fact that attributes are pairwise strongly dependent. This is a fairly complex problem and cannot be solved with the myopic heuristics.

TREE: Domain whose instances were generated from a decision tree with 6 internal nodes, each containing a different binary attribute. 5 random binary attributes were added to the description of instances. This problem should be easy for greedy decision tree learning algorithms while other approaches may have difficulties due to an inappropriate knowledge representation of the target concept.

PAR2: Parity problem with two significant binary attributes and 10 random binary attributes. 5% of randomly selected instances were labelled with wrong class. This problem is hard as there is a lot of attributes with equal score when evaluated with a myopic evaluation function, such as information gain.

PAR3: Same as PAR2 except that there were three significant attributes for the parity relation which makes the problem harder.

PAR4: Same as PAR2 except that there were four significant attributes for the parity relation which makes the problem the hardest among the parity problems used in our experiments.

Besides the artificial data sets above, we used also the following well known artificial data sets used by other authors (note that results of other authors can not be directly compared to our results as experimental conditions (training/testing splits) were not the same):

domain	#class	#atts.	#val/att.	# instances	maj.class (%)	entropy(bit)
INF1	3	12	2.0	200	36	1.58
INF2	3	21	2.0	200	36	1.58
TREE	2	11	2.0	200	57	0.99
PAR2	2	12	2.0	200	54	0.99
PAR3	2	13	2.0	200	54	0.99
PAR4	2	14	2.0	400	50	1.00
BOOL	2	6	2.0	640	67	0.91
LED	10	7	2.0	1000	11	3.33
KRK1	2	18	2.0	1000	67	0.92
KRK2	2	6	8.0	1000	67	0.92

Table 1 Basic description of artificial data sets

BOOL: Boolean function defined on 6 attributes with 10% of class noise (optimal recognition rate is 90%). The target function is:

$$Y = (X_1 \oplus X_2) \vee (X_3 \wedge X_4) \vee (X_5 \wedge X_6)$$

This data set was used by Smyth et al. (1990) and they report $67.2\pm1.7\%$ of the classification accuracy for naive Bayes, $82.5\pm1.1\%$ for backpropagation, and $85.9\pm0.9\%$ for their rule-based classifier.

LED: LED-digits problem with 10% of noise in attribute values. The optimal recognition rate is estimated to be 74%. Smyth et al. (1990) report $68.1\pm1.7\%$ of the classification accuracy for naive Bayes, 64.6 ± 3.5 for backpropagation, and 72.7 ± 1.3 for their rule-based classifier. This data set can be obtained from Irvine database (Murphy & Aha, 1991).

KRK1: The problem of legality of King-Rook-King chess endgame positions. The attributes describe the relevant relations between pieces, such as "same_rank" and "adjacent_file". Originally the data included five sets of 1000 examples (1000 for learning and 1000 for testing) and was used to test Inductive Logic Programming algorithms (Džeroski, 1991). The reported classification accuracy is $99.7\pm0.1\%$. We used only one set of 1000 examples (i.e. 700 instances for training).

KRK2: Same as KRK1 except that the only available attributes are the coordinates of pieces. The same data set was used by Mladenić (1993). The reported results are about 69% accuracy for her ATRIS system and 64% for Assistant.

The basic characteristics of the artificial data sets are listed in table 1. The results

domain	Assistant-I	Assistant-R	naive Bayes
INF1	88.8±4.0	88.2±4.2	89.8±3.7
INF2	59.7±10.	69.3±8.1	30.7±7.5
TREE	78.2±7.3	78.0±7.8	68.6±6.8
PAR2	69.9±5.7	94.6±2.7	54.6±4.2
PAR3	68.1±13.	94.6±3.0	55.6±5.7
PAR4	57.9±6.8	94.3±0.9	54.8±4.2
BOOL	89.0±1.7	89.0±1.7	67.5±2.7
LED	71.8±3.1	72.0±2.6	74.4±2.6
KRK1	98.4±2.0	98.4±2.0	91.7±2.0
KRK2	67.9±2.4	71.1±2.7	66.2±1.2

Table 2 Classification accuracy of the learning systems on artificial data sets

domain	Assistant-I	Assistant-R	naive Bayes
INF1	1.22±0.07	1.20±0.08	1.31±0.07
INF2	0.57±0.19	0.76±0.08	0.05±0.05
TREE	0.46±0.10	0.46±0.10	0.21±0.05
PAR2	0.28±0.11	0.81±0.04	0.07±0.02
PAR3	0.27±0.22	0.79±0.04	0.03±0.05
PAR4	0.11±0.10	0.79±0.02	0.06±0.02
BOOL	0.49±0.03	0.52±0.03	0.07±0.02
LED	2.12±0.07	2.14±0.07	2.33±0.06
KRK1	0.84±0.05	0.84±0.05	0.61±0.02
KRK2	0.14±0.04	0.19±0.04	-0.02±0.01

Table 3 Average information score of the learning systems on artificial data sets

of the learning algorithms Assistant-I and Assistant-R, as well as the naive Bayesian classifier are given in table 2 (classification accuracy) and table 3 (information score).

The results show that

- All classifiers perform well in a (relatively simple) domain with independent attributes (INF1).

- Both versions of Assistant perform well in the problem of the reconstruction of a decision tree (TREE), while the naive Bayesian classifier is significantly worse.

- Only Assistant-R is able to successfully solve the problems with strong dependencies between attributes (INF2, PAR2-4).

It is interesting that in the LED domain, the naive Bayesian classifier reaches the estimated upper bound of the classification accuracy. This suggests that all attributes should be considered for optimal classification in this domain. In the other three domains the performance of the naive Bayesian classifier is poor, due to the strong dependencies between attributes. The information score (see table 6) shows that the naive Bayesian classifier provides (on the average) no information in the BOOL and KRK2 domains.

The performance of the different variants of Assistant is almost the same, except for the KRK2 domain, where the performance of Assistant-I is poor (note that 67% of instances belong to the majority class). The performance of Assistant-R is slightly better

4.3   Real world data sets

We compared the performance of the algorithms also on several real world data sets. The following are problems from medicine:

- Data sets obtained from University Medical Center in Ljubljana, Slovenia: the problem of locating of primary tumour in patients with metastases (PRIM), the problem of predicting recurrence of breast cancer five years after the removal of the tumour (BREA), the problem of determining the type of the cancer in lymphography (LYMP), and diagnosis in rheumatology (RHEU).

- HEPA: prognostics of survival for patients suffering from hepatitis. The data was provided by Gail Gong from Carnegie-Mellon University.

We also compared the performance of the algorithms on the following non-medical real world data sets (SOYB, IRIS, and VOTE are obtained from the Irvine database (Murphy & Aha, 1991)):

SOYB: The famous soybean data set used by Michalski & Chilausky (1980).

IRIS: The well known Fisher's problem of determining the type of iris flower.

MESH3,MESH15: The problem of determining the number of elements for each of the edges of an object in the finite element mesh design problem (Dolšak & Muggleton, 1992). There are five objects for which experts have constructed appropriate meshes. In each of five experiments one object is used for testing and the other four for learning and the results are averaged. The results reported by Džeroski (1991) for various ILP systems are 12% classification accuracy for FOIL, 22% for mFOIL and 29% for GOLEM and the result reported by Pompe et al. (1993) is 28% for SFOIL. The description of the MESH problem is appropriate for ILP systems. For attribute learners only relations with arity 1 (i.e. attributes) can be used to describe the problem. Note that in this domain the training/testing splits are the same for all algorithms. The testing methodology is a special case of leave-one-out, therefore, the results in the tables for this problem have no standard deviations.

MESH3 contains the three basic attributes from the original database and ignores the relational description of objects. Therefore, in the MESH3 domain attribute learners are given less information than ILP learners.

MESH15 contains, besides the 3 original attributes, 12 attributes derived from the relational background knowledge. In this problem, attribute learners have advantage as they are already provided with additional attributes. The provided description of objects for ILP learners is actually more informative. In principle, the same attributes and a number of additional attributes could be derived by (extremely cleaver) ILP learners from the relational description of the background knowledge. However, this is a fairly complex task. Therefore attribute learners with MESH15 data set have better chances than ILP learners to reveal a good hypothesis.

VOTE: The voting records are from a session of the 1984 United States Congress. Smyth et al. (1990) report 88.9% of classification accuracy for the naive Bayesian classifier, 93.0% for backpropagation and 94.9% for their rule-based classifier.

CRIME: Problem from criminology: classification of criminals according to their education.

ELPI: The problem of discriminating electrons from pions in physics. The data set

domain	#class	#atts.	#val/att.	# instances	maj.class (%)	entropy(bit)
PRIM	22	17	2.2	339	25	3.89
BREA	2	10	2.7	288	80	0.73
LYMP	4	18	3.3	148	55	1.28
RHEU	6	32	9.1	355	66	1.73
HEPA	2	19	3.8	155	79	0.74
SOYB	15	35	2.9	630	15	3.62
IRIS	3	4	6.0	150	33	1.59
MESH3	13	3	7.0	278	26	3.02
MESH15	13	15	7.1	278	26	3.02
VOTE	2	16	2.0	435	61	0.96
CRIME	4	11	4.5	723	65	1.34
ELPI	2	18	16.6	500	50	1.00

Table 4 Basic description of the medical and non-medical real-world data sets

domain	Assistant-I	Assistant-R	naive Bayes
PRIM	42.2±6.1	40.1±5.8	51.0±3.8
BREA	78.1±5.1	78.5±4.5	79.2±5.2
LYMP	77.2±5.0	75.8±5.3	84.2±2.9
RHEU	67.0±4.7	66.2±4.3	67.2±5.0
HEPA	77.3±6.0	83.0±3.5	86.6±3.5
SOYB	91.1±4.2	89.0±3.2	89.1±2.1
IRIS	95.4±2.2	95.4±2.6	96.8±1.3
MESH3	32.7%	32.0%	33.5%
MESH15	41.0%	42.4%	34.5%
VOTE	96.3±1.6	95.3±1.3	90.3±2.0
CRIME	56.8±1.6	57.4±3.2	61.2±2.8
ELPI	84.3±3.2	81.1±2.2	84.7±3.0

Table 5 Classification accuracy of the learning systems on medical and non-medical real-world data sets

domain	Assistant-I	Assistant-R	naive Bayes
PRIM	1.19±0.13	1.08±0.10	1.64±0.25
BREA	0.06±0.07	0.05±0.07	0.08±0.06
LYMP	0.61±0.08	0.59±0.10	0.77±0.09
RHEU	0.44±0.09	0.41±0.09	0.54±0.07
HEPA	0.15±0.09	0.26±0.10	0.39±0.13
SOYB	3.10±0.10	3.01±0.10	3.26±0.10
IRIS	1.41±0.06	1.43±0.06	1.50±0.04
MESH3	0.57 bit	0.56 bit	0.65 bit
MESH15	0.74 bit	0.70 bit	0.56 bit
VOTE	0.84±0.03	0.83±0.03	0.75±0.04
CRIME	0.09±0.03	0.08±0.04	0.09±0.03
ELPI	0.63±0.05	0.57±0.03	0.70±0.05

Table 6 Average information score of the learning systems on medical and non-medical real-world data sets

The basic characteristics of the real-world data sets are given in table 4. The results of experiments on these data sets are provided in tables 5 and 6.

In medical data sets, attributes are typically independent. Therefore, it is not surprising that the naive Bayesian classifier shows clear advantage on these data sets (Kononenko, 1993).

The information score (table 6) for BREA data set indicates that no learning algorithm was able to solve this problem. This suggests that the attributes are not relevant.

Both versions of Assistant have similar performance, except in the HEPA domain where, Assistant-R has significantly better performance (99% confidence level using the two tailed t-test). A detailed analysis showed that in this problem RELIEFF discovered a significant interdependency between two attributes. These two attributes score poorly when considered independently. That is why Assistant-I was not able to discover this regularity in data.

On the other hand, other (redundant) attributes are available in HEPA that contain similar information as these two attributes together. This is the reason why the naive Bayesian classifier performs better. We tried to provide the naive Bayesian classifier with an additional attribute by joining the two dependent attributes. However, the performance was the same.

On SOYB, IRIS, CRIME, and ELPI data sets, all classifiers perform equally well. Note that the information score for the CRIME data set (table 6) indicates that the attributes are irrelevant to the class. On the VOTE data set the naive Bayesian classifier is the worst, while both versions of Assistant are comparable to the rule based classifier by Smyth et al. (1990).

The most interesting results appear in the MESH domains. Although attribute learners in MESH3 have less information than ILP systems, they all outperform the results by ILP systems as reported by Džeroski (1991) and Pompe et al. (1993). With 12 additional attributes in MESH15, the results of inductive learners are significantly improved. All inductive learning systems significantly outperform the naive Bayesian classifier in this problem.

A detailed analysis showed that this excellent result by both versions of Assistant is due to the use of the naive Bayesian formula to calculate the class probability distribution in "null" leaves (see section 3). Namely, for this problem it often happens that testing instances fall into a "null" leaf because there are no training instances that have the same values of significant attributes as the testing instances. The naive Bayesian classifier efficiently solves this problem.

5   Conclusions

Current inductive learning algorithms that use impurity functions for guiding the search are myopic due to greedy search. RELIEFF is an efficient heuristic estimator of attribute quality that is able to deal with data sets with dependent and independent attributes. The extensions of RELIEF enable it to deal with noisy, incomplete, and multi-class data sets. The interpretation of eq. (1) explains the interesting strong correlation of RELIEFF's estimates with other impurity functions when a large number (k) of nearest hits/misses is used.

The defficiency of myopic behavior of current inductive learning systems can be overcome by replacing the existing heuristic functions with RELIEFF. Assistant-R, a variant of top down induction of decision trees algorithms that uses RELIEFF for estimating the quality of the attributes, significantly outperforms other classifiers in domains with strong dependencies between attributes. The myopia of other inductive learners may cause them to overlook significant relations. While this can be easily demonstrated with artificial data sets, it was also shown in one real world problem HEPA. Here RELIEFF detected a significant interdependency between two attributes, that resulted in a significantly better result by Assistant-R than the result by Assistant-I.

Although RELIEFF may overcome the myopia, it is useless in Assistant-R when a representation change is required. In such cases, constructive induction should be applied. For example, in the KRK2 problem, Assistant-R achieves good result, which can not be further improved without constructive induction. Semi-naive Bayesian classifier (Kononenko, 1991) achied 86% classification accuracy. This algorithm explicitly searches for dependencies between attributes and costructs a new attribute when significant dependency is discovered. Such learning, when new attributes are constructed from existing ones, is called constructive induction. A good idea for guiding the constructive induction may be to use RELIEFF.

The naive Bayesian classifier has obvious advantage in domains with relatively independent attributes, such as medical diagnostic problems. In such domains, the naive Bayesian classifier is able to reliably estimate the conditional probabilities and is also able to use all attributes, i.e all available information. It would be interesting to appropriately combine the power of RELIEFF and the naive Bayesian classifier.

Current inductive logic programming (ILP) systems (Muggleton, 1992) are not able to use the attributes appropriately. This was demonstrated in the MESH3 domain where all attribute learners outperformed existing ILP systems. To enable ILP systems to deal with an attribute-value representation, a combination with the (semi) naive Bayesian classifier could be useful. On the other hand, current ILP systems also use greedy search techniques and the heuristics that guide the search are myopic. It would be interesting to use RELIEFF in ILP systems as well. Preliminary results using a version of RELIEFF in an ILP system are described by Pompe and Kononenko in this volume.

Acknowledgements

The use of m-estimate in equation (1) was proposed by Bojan Cestnik. We thank Matjaž Zwitter for the PRIM and BREA data sets, Milan Soklič for LYMP, Gail Gong for HEPA, Padhraic Smyth for BOOL and LED, Sašo Džeroski for KRK1 and MESII, Igor Mandič for ELPI, and Patrick Murphy and David Aha for the data sets from the Irvine database. We are grateful to our colleagues Sašo Džeroski, Matevž Kovačič, Matjaž Kukar, Uroš Pompe, Marko Robnik, and Tanja Urbančič for their comments on earlier drafts that significantly improved the paper. This work was supported by the Slovenian Ministry of Science and Technology.

References

1. Breiman L., Friedman J.H., Olshen R.A. & Stone C.J.: *Classification and Regression Trees*, Wadsworth International Group, 1984.

2. Cestnik B. : Estimating probabilities: A crucial task in machine learning, *Proc. European Conference on Artificial Intelligence 90*, Stockholm, August 1990, 147-149.

3. Cestnik B. & Bratko I.: On estimating probabilities in tree pruning, *Proc. European Working Session on Learning*, (Porto, March 1991), Y.Kodratoff (ed.), Springer Verlag, 1991, pp.138-150.

4. Cestnik, B., Kononenko, I. & Bratko, I.: ASSISTANT 86: A knowledge elicitation tool for sophisticated users, in: *Progress in Machine Learning* (Eds. I. Bratko & N. Lavrač), Sigma Press, Wilmslow, England, 1987.

5. Dietterich T.G. & Shavlik J.W. (eds.): *Readings in machine learning*, Morgan Kaufmann, 1990.

6. Dolšak, B. & Muggleton, S.: The application of inductive logic programming to finite element mesh design, in: *Inductive Logic Programming* (Ed. S. Muggleton), Academic Press, 1992.

7. Džeroski S.: Handling noise in inductive logic programming, M.SC. Thesis, University of Ljubljana, Faculty of electrical engineering & computer science, Ljubljana, Slovenia, 1991.

8. Fayyad U.M.: On the induction of decision trees for multiple concept learning, Ph.D. Thesis, The University of Michigan, 1991.

9. Fayyad U.M. & Irani K.B.: The attribute selection problem in decision tree generation, *Proc. AAAI-92* (San Jose, CA, July 1992), MIT Press, 1992.

10. Hunt E., Martin J & Stone P.: *Experiments in Induction*, New York, Academic Press, 1966.

11. Kira K. & Rendell L. (a): A practical approach to feature selection, *Proc. Intern. Conf. on Machine Learning* (Aberdeen, July 1992), (Eds. D.Sleeman & P.Edwards), Morgan Kaufmann, 1992, 249-256.

12. Kira K. & Rendell L. (b): The feature selection problem: traditional methods and new algorithm, *Proc. AAAI'92* (San Jose, CA, July 1992), MIT Press, 1992.

13. Kononenko I.: Semi-naive Bayesian classifier, *Proc. European Working Session on Learning*, (Ed. Y.Kodratoff), Springer Verlag, 1991, 206-219.

14. Kononenko I.: Inductive and Bayesian learning in medical diagnosis. *Applied Artificial Intelligence*, 7 (1993), 317-337.

15. Kononenko I.: Estimating attributes: Analysis and extensions of RELIEF. *Proc. European Conf. on Machine Learning* (Catania, April 1994) (Ed. L. De Raedt & F.Bergadano), Springer Verlag, 1994, 171-182.

16. Kononenko, I. & Bratko, I.: Information based evaluation criterion for classifier's performance. *Machine Learning*, 6 (1991), 67-80.

17. Mantaras R.L.: ID3 Revisited: A distance based criterion for attribute selection, *Proc. Int. Symp. Methodologies for Intelligent Systems*, Charlotte, North Carolina, U.S.A., Oct. 1989.

18. Michalski, R.S. & Chilausky, R.L.: Learning by being told and learning from examples: An experimental comparison of the two methods of knowledge acquisition in the context of developing an expert system for soybean disease diagnosis. *International Journal of Policy Analysis and Information Systems*, 4 (1980), 125-161.

19. Michalski, R.S. & Tecuci G. (eds.): *Machine learning: A multistrategy approach*, Vol. IV, Morgan Kaufmann, 1994.

20. Mladenič D.: Combinatorial optimization in inductive concept learning, *Proc. 10th Intern. Conf. on Machine Learning* (Amherst, June 1993), Morgan Kaufmann, 1993, 205-211.

21. Muggleton S. (ed.): *Inductive Logic Programming*, Academic Press, 1992.

22. Murphy, P. M., & Aha, D. W.: *UCI Repository of machine learning databases* [Machine-readable data repository]. Irvine, CA: University of California, Department of Information and Computer Science, 1991.

23. Niblett, T. & Bratko, I.: Learning decision rules in noisy domains, *Proc. Expert Systems 86*, Brighton, UK, December 1986.

24. U.Pompe, M.Kovačič & I.Kononenko: SFOIL: Stochastic approach to inductive logic programming. *Proc. Slovenian Conf. on Electrical Engineering and Computer Science*, Portorož, Slovenia, Sept. 1993, 189-192.

25. Quinlan R.: Induction of decision trees, *Machine Learning*, 1 (1986), 81-106.

26. Smyth P. & Goodman R.M.: Rule induction using information theory, in: *Knowledge Discovery in Databases* (Eds. G.Piatetsky-Shapiro & W.Frawley), MIT Press, 1990.

27. Smyth P., Goodman R.M. & Higgins C.: A hybrid Rule-based Bayesian Classifier, *Proc. European Conf. on Artificial Intelligence*, Stockholm, August, 1990, 610-615.

REPRESENTING EXAMPLES BY EXAMPLES

M. Bohanec
J. Stefan Institute, Ljubljana, Slovenia
and
I. Bratko
University of Ljubljana, Ljubljana, Slovenia

Abstract

The problem addressed in this paper is: given a set of examples representing some target concept, construct a minimal subset consisting of the most representative examples from which the original set could be easily and accurately reconstructed based on domain-specific background knowledge about the class of target concepts. Two methods are presented that assume two properties of the target concept class, respectively: monotonicity and linearity. Their performance is empirically assessed on about 500 sets of examples from 40 real-life decision problems, developed with expert system shells for multi-attribute decision support DECMAK and DEX. The results indicate that the first method is of practical interest in typical decision domains. It is simpler, more efficient, and generates comprehensible representations. In contrast, although the second method typically generates sets of representatives that are of similar size or even smaller, its results seem to be much less comprehensible.

1 Introduction

In attribute-based inductive learning, the goal is to transform a set of examples into a suitable representation that accurately and comprehensibly represents the target concept. Typical representation formalisms are rules and decision trees.

In this paper we investigate an approach to represent the target concept in the basic formalism, i.e., by *examples* themselves. The approach is based on the assumption that a substantial portion of the examples in the initial learning set are redundant in the sense that they could be easily reconstructed from other examples, taking into account domain-specific knowledge about the class of target concepts. Therefore, in order to represent the target concept, it would be sufficient to identify a subset consisting of the most relevant (*representative*) examples from which all the original examples (and possibly the target concept itself) could be fully and accurately reconstructed. For the reason of comprehensibility, the subset of representatives should be as small as possible.

The critical part of this approach is the reconstruction process. Instead of an explicit representation of the target concept, people are given representative examples in the hope that they will accurately reconstruct the concept themselves. But which mechanisms and rules do people use in doing this, and which are the most effective? Which properties of the target concept are easy to reconstruct, and which are difficult? Is the minimal subset sufficient for an accurate reconstruction, or do we need additional (redundant) examples? We still cannot generally answer to these difficult questions. However, we can focus on specific problem domains with distinctive and well-understood properties of target concepts, and try to find the answers there.

In this paper we concentrate on problem domains that occur in qualitative multi-attribute decision making. Specifically, we consider the so-called *utility functions* that are defined by the decision maker in order to express their preferences related to various options (alternatives) involved in the decision-making process.

In the following section, we introduce the concept of utility functions, illustrate their basic properties, and motivate the research. In sections 3 and 4 we present two methods for the construction of a minimal set of representative examples, respectively based on two assumptions about the properties of the target utility function: (1) monotonicity and (2) linearity. For these methods we show examples of generated representations, illustrate the reverse process of reconstructing the original function and present a statistical analysis of their performance on 534 different utility functions. We conclude by a critical analysis and comparison of the methods, particularly with respect to the size and comprehensibility of generated representations.

2   Utility Functions

In general, a *decision-making process* can be described as the process of selecting a particular option (alternative) from a set of possible ones so as to best satisfy the aims or goals of the decision maker. In decision theory [14], this is called the option with the highest *utility*.

A number of methods and computer-based systems have been developed to support people in making complex real-world decisions [16, 23]. A well-known approach is based on the development of *multi-attribute decision models* [8]. The basic principle is a hierarchical decomposition of the decision problem into specific attributes [21]. Options are first evaluated independently with respect to each attribute. The total utility of options is then obtained by some aggregation procedure and used as a basis for the final decision.

A typical multi-attribute decision model consists of two components. First, there is a *hierarchy of attributes* that represents the global structure of the problem domain. In machine learning, similar hierarchies are used, for example, in signature table learning [5] and structured induction [22]. The second component consists of one or more *utility functions* that define a bottom-up mapping of attributes in the hierarchy and thus determine the aggregation procedure.

In this paper we focus on utility functions. More specifically, since different methods and computer programs treat them differently, we consider utility functions in the form as they are used in DECMAK [6] and DEX [7], two expert system shells for qualitative multi-attribute decision support. In both systems, the functions are defined by *examples*, which are provided by the decision maker.

	SW	*HW*	*SYSTEM*
1.	unacc	unacc	unacc
2.	unacc	acc	unacc
3.	unacc	good	unacc
4.	unacc	exc	unacc
5.	acc	unacc	unacc
6.	acc	acc	acc
7.	acc	good	acc
8.	acc	exc	good
9.	good	unacc	unacc
10.	good	acc	good
11.	good	good	good
12.	good	exc	good
13.	exc	unacc	unacc
14.	exc	acc	good
15.	exc	good	exc
16.	exc	exc	exc

Table 1: A utility function for the evaluation of a computer *SYSTEM* based on the evaluations of its software (*SW*) and hardware (*HW*)

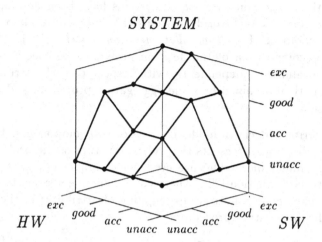

Figure 1: Graphical representation of the *SYSTEM* utility function

A small, but typical set of examples is presented in Table 1. It defines a utility function that evaluates the quality of a computer system based on the lower-level evaluations of its software and hardware. Each line in the table represents a single example that was provided by the decision maker. Each of the two attributes, *SW* and *HW*, and the class *SYSTEM* can take one of the four values: *unac*ceptable, *acc*eptable, *good*, or *exc*ellent. There are 16 possible combinations of the values of *SW* and *HW*; they are all defined in Table 1, so the function is completely defined. In general, the definition need not be *complete*. On the other hand, it is required to be *consistent* so that at most one class is defined for each combination of attribute values.

Majority of utility functions exhibit some distinct properties. First, the sets of attribute and class values are often *ordered* by preference: the higher value is the more preferred. The result is that the corresponding utility function is *monotone*: when values of one or more attributes increase, the function increases or remains constant. For the function *SYSTEM* defined above, this is illustrated by a graphical representation in which the examples are shown as points in three dimensions (Figure 1). Another property is that functions or at least their parts are *linear*, so they can be sufficiently well approximated by a linear function. (The term 'linear' is for our domains meant under the encoding of the possible attribute and class values by successive integers. The meaning of 'sufficiently well approximated' is clarified in section 4.) The function in Figure 1 is linear in the region where *HW* takes values from *acc* to *exc*.

Since 1981, 40 decision models have been developed by DECMAK and DEX in various real-world decision problems, such as personnel management, technology evaluation, investment planning and performance evaluation of enterprises [24]. Each model usually contains ten or more utility functions, so the total number of utility functions defined so far is 534. Among them, as many as 92.51% are monotone, but only 13.86% are totally linear. However, in average about 73% of examples defining a function can be linearly approximated. In general, utility functions are relatively small, being defined by about 24 examples in average. The number of attributes is typically between 2 and 5, and the number of attribute values and classes is between 2 and 8.

Although the utility functions consist of a relatively small number of examples and can be effectively obtained from the decision maker [18], they are often too complex to be easily reviewed and understood by people when represented in their original form (e.g., Table 1). Therefore, a more compact and comprehensible representation is needed. An obvious solution would be to use a machine learning algorithm that develops decision trees [17] or rules [15, 9]. We tried that [19], but the results were not completely satisfactory. Decision trees tended to grow large in comparison to the size of the learning set. Rules performed considerably better in terms of size and comprehensibility; however, a notable proportion of decision makers found them difficult to read and understand since they were so different from the more familiar examples. In other words, the different representations required the decision makers to learn different representation formalisms, which was often difficult to achieve.

This motivated us to try an alternative approach in which utility functions are represented by examples themselves. The representation is made more compact (and possibly more comprehensible) by eliminating redundant examples. What remains is a subset of representative examples from which the eliminated ones can be uniquely reconstructed assuming a specific property of the utility function. In the following two sections we present two methods that construct representatives, respectively based on the assumptions of monotonicity and linearity.

3   Representing Monotone Functions

Let $u : \mathcal{X} \longrightarrow \mathcal{C}$ be a utility function, where $\mathcal{C} = \{1, 2, \ldots, m\}$ is a preferentially ordered set of classes and $\mathcal{X} = X_1 \times X_2 \times \cdots \times X_a$, where each X_i is a discrete and possibly ordered attribute. The function is partially defined by a set of examples $\mathcal{E} = \{(\mathbf{x}, c), \mathbf{x} \in \mathcal{X}, c \in \mathcal{C}\}$ so that

$$(\mathbf{x}, c) \in \mathcal{E} \implies u(\mathbf{x}) = c$$

In this section we assume that the function is monotone. A function is *monotone* if for each pair of examples $(\mathbf{x}_1, c_1), (\mathbf{x}_2, c_2) \in \mathcal{E}$: if $\mathbf{x}_1 \succeq \mathbf{x}_2$ then $c_1 \geq c_2$. The relation '$\succeq$' that compares two attribute-value vectors is defined according to the (possible) preference order of attributes. More precisely, let $\mathbf{x}_1 = \langle p_1, p_2, \ldots, p_a \rangle$ and

$\mathbf{x}_2 = \langle q_1, q_2, \ldots, q_a \rangle$. Then $\mathbf{x}_1 \succeq \mathbf{x}_2$ whenever $p_i \geq q_i$ for all preferentially ordered attributes X_i, and $p_j = q_j$ for all unordered attributes X_j. Note that this relation may not be defined for some pairs of vectors.

First, let us ask which examples in $\mathcal{E}$ are redundant in the sense that they could be reconstructed from other examples knowing that the function is monotone. Take an example $(\mathbf{x}_0, c_0) \in \mathcal{E}$. According to the relation '$\succeq$', this example defines two subsets of the attribute-value space:

$$\mathcal{X}^+(\mathbf{x}_0) = \{\mathbf{x} \in \mathcal{X} : \mathbf{x} \succeq \mathbf{x}_0\}$$

$$\mathcal{X}^-(\mathbf{x}_0) = \{\mathbf{x} \in \mathcal{X} : \mathbf{x}_0 \succeq \mathbf{x}\}$$

These two subsets are illustrated in Figure 2 for the case of two ordered attributes.

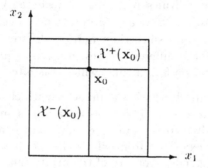

Figure 2: Subspaces $\mathcal{X}^+(\mathbf{x}_0)$ and $\mathcal{X}^-(\mathbf{x}_0)$ for two ordered attributes, x_1 and x_2

It follows from the definition of monotonicity that $u(\mathbf{x}) \succeq c_0$ for each $\mathbf{x} \in \mathcal{X}^+(\mathbf{x}_0)$. That is, $(\mathbf{x}_0, c_0)$ determines the lower bound of the function in $\mathcal{X}^+(\mathbf{x}_0)$. In this area, all the remaining examples with the class of c_0 can be reconstructed from the lower bound, so they are redundant.

Consequently, the set of representative examples can be constructed by the following method, adapted from [4]. For each fixed class c, all the redundant examples are first discarded, resulting in a set $\mathcal{M}_c^+$ that consists of examples having the class c and whose attribute-value vectors can not be compared with respect to '$\succeq$':

$$\mathcal{M}_c^+ = \{(\mathbf{x}_0, c) \in \mathcal{E} : \forall(\mathbf{x}, c) \in \mathcal{E} : \mathbf{x} \succeq \mathbf{x}_0\}$$

The set of all representatives $\mathcal{M}^+$ is finally obtained as the union of $\mathcal{M}_c^+$ for all classes, except for $c = 1$, which is always the lower bound of the whole function and is therefore redundant as well.

Given $\mathcal{M}^+$, it is easy to reconstruct the original function, provided that the function was monotone and completely defined by the examples. For all $(\mathbf{x}_0, c) \in \mathcal{M}^+$ it is obvious that $u(\mathbf{x}_0) = c$. When for a given $\mathbf{x}_0$ there is no corresponding example in $\mathcal{M}^+$, then this example has been eliminated because its class was equal to the lower bound of u at $\mathbf{x}_0$. This bound can be, however, determined from the representatives so that

$$u(\mathbf{x}_0) = \max_{(\mathbf{x},c) \in \mathcal{M}^+ : \mathbf{x}_0 \succeq \mathbf{x}} c$$

In determining this, it is possible that none of the examples $(\mathbf{x}, c) \in \mathcal{M}^+$ satisfies $\mathbf{x}_0 \succeq \mathbf{x}$. In this case $u(\mathbf{x}_0) = 1$, since 1 is the only value whose representatives have been excluded from $\mathcal{M}^+$.

It is worth noting that when the original function is incomplete (partially defined), the reconstruction process involves generalization, i.e., assigns classes also to attribute-value vectors that were originally undefined. In this case, the reconstruction procedure assigns the lower bound.

In addition to $\mathcal{M}^+$, another set of representatives $\mathcal{M}^-$ can be constructed similarly by determining the upper bounds of u in each subspace $\mathcal{X}^-(\mathbf{x}_0)$, that is

$$\mathcal{M}_c^- = \{(\mathbf{x}_0, c) \in \mathcal{E} : \forall (\mathbf{x}, c) \in \mathcal{E} : \mathbf{x}_0 \succeq \mathbf{x}\}$$

and

$$\mathcal{M}^- = \bigcup_{c=1}^{m-1} \mathcal{M}_c^-$$

A monotone and completely defined function can also be fully reconstructed from $\mathcal{M}^-$ with a similar procedure.

In order to treat utility functions that are initially non-monotone, the construction procedures have to be slightly extended. In non-monotone functions, there exist examples $\mathbf{x} \succeq \mathbf{x}_0$ whose class is lower than the assumed lower bound of u in $\mathcal{X}^+(\mathbf{x}_0)$, or greater than the upper bound of u in $\mathcal{X}^-(\mathbf{x}_0)$. Before or during the construction, such examples can be eliminated from $\mathcal{E}$, but added to the final representation and explicitly labeled as exceptions that violate monotonicity.

As an example, consider the utility function from Table 1 and construct the corresponding $\mathcal{M}^+$ and $\mathcal{M}^-$. The results are presented in Table 2. Instead of original 16 examples, the function is now represented by only 4 and 5 representatives, respectively. It can be completely reconstructed from each of the subsets. The reconstruction from $\mathcal{M}^+$ can be interpreted as follows. The evaluation of a computer system is unacceptable until both hardware and software are acceptable (example 6). Then, the system is acceptable, too. The value of *SYSTEM* increases to good when *HW* increases to

	SW	*HW*	*SYSTEM*

$\mathcal{M}^+$:

	SW	*HW*	*SYSTEM*
6.	acc	acc	acc
8.	acc	exc	good
10.	good	acc	good
15.	exc	good	exc

$\mathcal{M}^-$:

	SW	*HW*	*SYSTEM*
14.	exc	acc	good
12.	good	exc	good
7.	acc	good	acc
13.	exc	unacc	unacc
4.	unacc	exc	unacc

Table 2: Representative examples for *SYSTEM* assuming monotonicity

excellent (8) or *SW* to good (10). The evaluation is excellent only when *SW* is excellent and *HW* is at least good (15). A similar—but 'backwards'—reconstruction follows also from $\mathcal{M}^-$.

We tested the methods presented in this section on 534 utility functions obtained so far. In average, only about one fifth (21.36%) of the original examples were included in the sets $\mathcal{M}^+$ or $\mathcal{M}^-$, thus a considerable reduction in the representation size was achieved. However, the reduction was function-dependent, so the proportion of representatives largely varied between 3.33% and 77.78%. As expected, this proportion was the highest with non-monotone functions, but also tended to be rather high with functions that were linear or close to that. Since some functions were non-monotone, there were also 212 exceptions generated in total (less than half an exception per function in average).

It is interesting to compare the sets $\mathcal{M}^+$ and $\mathcal{M}^-$. Although they could be considered equivalent with respect to the reconstruction (except for the generalization of previously undefined examples), $\mathcal{M}^+$ tends to be smaller than $\mathcal{M}^-$ in average, so it is usually preferred. Based on a review of generated representations it also seems that it is slightly easier for people to perform the 'forward' reconstruction from $\mathcal{M}^+$ than the 'backward' one from $\mathcal{M}^-$. However, this is a subjective conjecture that needs additional justification.

4   Representing Linear Functions

In the context of linearity, we interpret utility functions geometrically as (discrete) points in multidimensional space (such as in Figure 1). Most importantly, we replace all the linguistical attribute values with their ordinal numbers and hereafter assume that $\mathbf{x} \in \mathcal{R}^a$. A utility function $u(\mathbf{x})$ can then be approximated by a linear function $h_*(\mathbf{x})$ such that

$$E = \max_{\mathbf{x} \in \mathcal{X}} |u(\mathbf{x}) - h_*(\mathbf{x})|$$

is minimal.

The representatives of u are those points from which h_* can be uniquely constructed. Conversely, once h_* is known, it is possible to reconstruct those points of u that were previously sufficiently close to h_*. Whenever $|u(\mathbf{x}) - h_*(\mathbf{x})| < 0.5$, such a reconstruction is possible by rounding h_* to the nearest integer. Otherwise, the reconstruction is wrong, so the point $\mathbf{x}$ must be explicitly represented as an exception to h_*.

In order to make this schema work, we have to establish a relationship between the set of representatives and h_*. One of Chebyshev's theorems [11], adapted to the notation and terminology used here, does exactly this:

Theorem (Chebyshev) *The function h_* of a attributes is determined by points*

$$\mathcal{L} = \{\mathbf{x}_1, \mathbf{x}_2, \ldots, \mathbf{x}_{a+2}\}$$

such that

1. $|u(\mathbf{x}_i) - h_*(\mathbf{x}_i)| = E$, $i = 1, 2, \ldots, a + 2$, and

2. *the elements of the sequence* $(u(\mathbf{x}_i) - h_*(\mathbf{x}_i))$, $i = 1, 2, \ldots, a + 2$, *have alternating signs.*

Therefore, h_* can be represented by $a + 2$ well-chosen points. The difference $|u(\mathbf{x}_i) - h_*(\mathbf{x}_i)|$ should be equal to E in all these points and at the same time greater than the difference in the remaining points. One half of the points must lie above, and the other half below h_*. More precisely, the number of points lying above and below h_* may differ by 1 at the most. We will denote these 'halves' by $\mathcal{L}^+$ and $\mathcal{L}^-$, respectively.

The algorithm for constructing $\mathcal{L}^+$ and $\mathcal{L}^-$ from a larger initial set of points is known and utilized in numerical analysis [13], so we exclude it from this presentation. Instead, we again take the *SYSTEM* utility function from Table 1 and develop its representation in the form of $\mathcal{L}^+$ and $\mathcal{L}^-$. The obtained four representatives and two exceptions are shown in Table 3. The exceptions were generated because the function was not totally linear.

	SW	HW	SYSTEM

$\mathcal{L}^-$:

	SW	HW	SYSTEM
4.	unacc	exc	unacc
13.	exc	unacc	unacc

$\mathcal{L}^+$:

	SW	HW	SYSTEM
1.	unacc	unacc	unacc
10.	good	acc	good

Exceptions:

	SW	HW	SYSTEM
6.	acc	acc	acc
15.	exc	good	exc

Table 3: Representative examples for *SYSTEM* assuming linearity

Apart from the two exceptions, the representation in Table 3 is as small as the first one in Table 2. One might presume that they are equally comprehensible as well. However, in the following we demonstrate that the reconstruction of u from the sets $\mathcal{L}^+$ and $\mathcal{L}^-$, although correct and complete, is considerably more difficult than the reconstruction from $\mathcal{M}^+$. To facilitate the reconstruction, let us first temporarily ignore the exceptions and present the examples from $\mathcal{L}^+$ and $\mathcal{L}^-$ in the form as shown in Table 4.

	SYSTEM	HW=unacc $x_2 = 1$	acc 2	good 3	exc 4
SW=unacc $x_1 = 1$		unacc 1			unacc 1
acc 2					
good 3			good 3		
exc 4		unacc 1			

Table 4: A tabular representation of $\mathcal{L}^+$ and $\mathcal{L}^-$ for *SYSTEM*

The next step is to construct

$$h_*(\mathbf{x}) = w_1 x_1 + w_2 x_2 + w_0$$

from $\mathcal{L}^+$ and $\mathcal{L}^-$ so as to satisfy the Chebyshev's theorem. Denote by $\mathbf{e}_1^+$ and $\mathbf{e}_2^+$ the two examples from $\mathcal{L}^+$ so that

$$e_1^+ = \langle x_{11}^+, x_{21}^+, c_1^+ \rangle = \langle 1, 4, 1 \rangle$$

$$e_2^+ = \langle x_{12}^+, x_{22}^+, c_2^+ \rangle = \langle 4, 1, 1 \rangle$$

Similarly, e_1^- and e_2^- are the points from $\mathcal{L}^-$:

$$e_1^- = \langle x_{11}^-, x_{21}^-, c_1^- \rangle = \langle 1, 1, 1 \rangle$$

$$e_2^- = \langle x_{12}^-, x_{22}^-, c_2^- \rangle = \langle 3, 2, 3 \rangle$$

Then, h_* must be parallel to the vectors $e_2^+ - e_1^+$ and $e_2^- - e_1^-$. In addition, the distance between h_* and these two vectors must be equal, so h_* must cross the point

$$t = (e_1^- - e_1^+)/2 = \langle t_1, t_2, t_c \rangle = \langle 0, -1.5, 0 \rangle$$

From these requirements it follows, using elementary vector algebra, that w_0, w_1 and w_2 can be obtained by solving:

$$\left\| \begin{array}{c} c_1^+ + t_c \\ c_2^+ + t_c \\ c_2^- - t_c \end{array} \right\| = w_1 \left\| \begin{array}{c} x_{11}^+ + t_1 \\ x_{12}^+ + t_1 \\ x_{12}^- - t_1 \end{array} \right\| + w_2 \left\| \begin{array}{c} x_{21}^+ + t_2 \\ x_{22}^+ + t_2 \\ x_{22}^- - t_2 \end{array} \right\| + w_0 \left\| \begin{array}{c} 1 \\ 1 \\ 1 \end{array} \right\|$$

The resulting function in our case is

$$h_*(\mathbf{x}) = 0.67\, x_1 + 0.67\, x_2 - 1.33$$

and tabulates as follows:

h_*	$x_2 = 1$	2	3	4
$x_1 = 1$	0.00	0.67	1.33	2.00
2	0.67	1.33	2.00	2.67
3	1.33	2.00	2.67	3.33
4	2.00	2.67	3.33	4.00

Finally, the original function u is reconstructed by rounding the above values to the nearest integer, except for the examples given in Table 3 (including exceptions) which are directly inserted into the final table (these values are denoted by '*'):

u	$x_2 = 1$	2	3	4
$x_1 = 1$	*1	1	1	*1
2	1	*2	2	3
3	1	*3	3	3
4	*1	3	*4	4

To summarize, the method considered in this section is attractive because of its ability to represent any linear utility function by only $a + 2$ representative examples. It is also interesting from the viewpoint of multi-attribute decision making since the majority of methods in this field rely on linear utility functions. Unfortunately, the representative example sets generated by this method seem to be highly incomprehensible. Although they are small, they require a difficult reconstruction procedure. The involved calculations are far from obvious and quite hard to be carried out by a human without additional tools. This is in sharp contrast with the representations used in section 3, where the reconstruction is easy and based only on a simple concept of preference orders.

5   Related Work

Learning and classification of monotonic ordinal concepts was studied by Ben-David *et al* [4]. They developed an algorithm that constructs a CISE, consistent and irredundant set of examples, based on the assumption of monotonicity and a 'most conservative classification principle'. The algorithm is capable of handling missing values: *don't knows* and *don't cares*. In section 3, we used an adapted version of this algorithm.

In artificial intelligence, there are several approaches based on the idea of representing knowledge by examples, most notably *instance-based learning* (IBL) [1] and *case-based reasoning* (CBR) [2, 20]. IBL algorithms are derived from the nearest neighbor pattern classifier [10] and based on the assumption that similar instances have similar classifications. They involve three components: (1) *similarity function* that computes the similarity between a training instance and the instances in the concept description, (2) *classification function* that classifies a training instance, and (3) *concept description updater* that decides which instances to include in the concept description. CBR systems are typically even more elaborate; they also modify cases and use parts of cases during problem solving. In comparison with our approach, in which we emphasize the comprehensibility of concept descriptions and require a complete reconstruction of the original concept, both IBL and CBR highlight other performance criteria, such as generality, classification accuracy, learning rate, and storage requirements [1].

Related approaches were—somewhat earlier—studied also in the field of clustering. Anderberg [3] utilized ordinal properties by mapping them into an interval level scale. There is a class of clustering algorithms [12] that associate each cluster with a *leader*, a typical representative of that cluster. The set of leaders is used for both the representation of the problem domain and classification of new objects, using a convenient distance measure.

6 Conclusion

Representing a possibly large set of examples with a small subset of the most relevant representative examples is interesting particularly because (1) it may considerably reduce the representation size and thus improve its comprehensibility, and (2) the form of the representation remains unchanged. However, finding such a representation involves two assumptions related to the problem domain and performance of people in that domain. First, a specific property must be assumed (or possibly discovered) that discriminates between relevant and redundant examples. Second, it is assumed that people, given the set of representatives, will be able to apply that property and reconstruct the original set with sufficient ease.

In this paper we focused on a specific class of problem domains that occur in practical decision making, and considered methods for the construction and reconstruction of representatives based on two assumptions: monotonicity and linearity. Both types of methods achieved considerable improvement in terms of representation size; in measurements performed on 534 real sets of examples, the representations were in average reduced to only about one fifth of the original size.

The comparison of the two methods reveals some interesting differences. The method based on linearity is capable of representing any linear utility function by only $a + 2$ examples, where a is the number of attributes. However, since typical utility functions are not totally linear, exceptions are added to the final set, so the representation generated by both methods are approximately of the same size in average. The greatest difference in favor of the linearity-based method typically occurs with functions that are linear and monotone at the same time. In terms of comprehensibility, on the other hand, the monotonicity-based method is favorable and of particular practical interest. The reconstruction of the initial set of examples, involving comparison of preferentially ordered values, is easy and seems to be quite close to human reasoning. The reconstruction in the linear case is considerably less obvious. In other words, this is a good example of representations that are approximately of the same form and size, but substantially differ in terms of comprehensibility.

In future, we plan to justify the approach in realistic decision-making processes, particularly from the viewpoint of comprehensibility in comparison with other representation formalisms. We also wish to generalize the approach by taking into account some other properties of utility functions, for example symmetricity, and by considering other — deterministic and noisy — problem domains with different properties.

Acknowledgments

This work was supported by the Ministry of Science and Technology of the Republic of Slovenia.

References

[1] Aha, D.W., Kibler, D., Albert, M.K.: Instance-based learning algorithms. *Machine Learning* 6, 1991, 37–66.

[2] Aha, D.W.: Case-based learning algorithms. In: Bareiss, R. (Ed.) *Proceedings: Case-Based Reasoning Workshop*, San Mateo, California. Morgan Kaufman, Washington, 1991.

[3] Anderberg, M.R.: *Cluster Analysis for Applications.* Academic Press, New York, 1973.

[4] Ben-David, A., Sterling, L., Pao, Y.-H.: Learning and classification of monotonic ordinal concepts. *Computational Intelligence* 5, 1989, 45–49.

[5] Biermann, A.W., Fairfield, J.R.C., Beres, T.R.: Signature table systems and learning. *IEEE Transactions on Systems, Man and Cybernetics* SMC-12(5), 1982.

[6] Bohanec, M., Rajkovič, V.: An expert system approach to multi-attribute decision making. In: Hamza, M.H. (Ed.) *Proceedings of the IASTED Conference on Expert Systems.* Acta Press, Anaheim, 1987.

[7] Bohanec, M., Rajkovič, V.: DEX: An expert system shell for decision support. *Sistemica* 1, 1990, 145–157.

[8] Chankong, V., Haimes, Y.Y.: *Multiattribute Decision Making: Theory and Methodology.* North-Holland, Amsterdam, 1983.

[9] Clark, P.E., Niblett, T.: The CN2 induction algorithm. *Machine Learning* 3, 1989, 261–284.

[10] Cover, T.M., Hart, P.E.: Nearest neighbor pattern classification. *Institute of Electrical and Electronics Engineers Transactions on Information Theory* 13, 1967, 21-27.

[11] Daugavet, I.K.: *Introduction to the Theory of Function Approximation* (in Russian). ILU, Leningrad, 1977.

[12] Diday, E.: *Optimisation en classification automatique.* INRIA, Rocquencourt, 1979.

[13] Fox, L., Mayers, D.F.: *Computing Methods for Scientists and Engineers.* Clarendon Press, Oxford, 1968.

[14] French, S.: *Decision Theory: An Introduction to the Mathematics of Rationality.* Wiley, New York, 1986.

[15] Michalski, R.S., Mozetič, I., Hong, J., Lavrač, N.: The multi-purpose incremental learning system AQ15 and its testing applications to three medical domains. AAAI 86, Philadelphia, 1986.

[16] O'Keefe, R.M.: The evaluation of decision-aiding systems: Guidelines and methods. *Information & Management* 17, 1989, 217–226.

[17] Quinlan, J.R.: Induction of decision trees. *Machine Learning* 1, 1986, 81–106.

[18] Rajkovič, V., Bohanec. M., Batagelj. V.: Knowledge engineering techniques for utility identification. *Acta Psychologica* 68, 1988, 271–286.

[19] Rajkovič, V., Bohanec, M.: Decision support by knowledge explanation. In Sol, H.G., Vecsenyi, J. (Eds.), *Environments for Supporting Decision Processes*. North-Holland, Amsterdam, 1991, 47–57.

[20] Richter, M.M., Wess, S., Althoff, K.-D., Maurer, F. (Eds.): *First European Workshop on Case-Based Reasoning EWCBR-93*. SEKI Report SR-93-12, Universität Kaiserslautern, Germany, 1993.

[21] Saaty, T.L.: *Multicriteria Decision Making: The Analytic Hierarchy Process*. RWS Publications, Pittsburgh, 1990.

[22] Shapiro, A.: *Structured Induction of Expert Systems*. Addison-Wesley, Reading, 1987.

[23] Stewart, T.J.: A critical survey on the status of multiple criteria decision making theory and practice. *Omega* 20(5/6), 1992, 569–586.

[24] Urbančič, T., Kononenko. I., Križman, V.: *Review of Applications by Ljubljana Artificial Intelligence Laboratories*. Institut Jožef Stefan, Report DP-6218, Ljubljana, 1991.

GREEN COFFEE GRADING USING FUZZY CLASSIFICATION

F. Suggi Liverani

Illycaffè s.p.a., Trieste, Italy

1. Green coffee, market and product evaluation

Green coffee represents an important part of the international commodity market. Factors determining product characteristics, and consequently product prices, are: the country of origin, the processing method, crop, presence of defects, shape, grading, bean color and flavor, aroma and body that are very important for producing high quality espresso coffee.

Various standards of product classification are applied by trade organizations in both producing countries and consumer countries; in many cases, such standards have been partially embodied in national legislation.

Methods of classifying coffee characteristics differ from country to country. Suffice it to mention the New York Coffee and Sugar Exchange system which is accepted in Brazil, and which sets seven standard types on the basis of the quantity of imperfections found in a one-pound sample. A similar system developed by French legislation[1] is accepted in French-speaking producing countries and since 1964, the International Standards Organization (ISO) has endeavored to establish precise norms on quality definition[2].

---

[1] Decree 65-763 September 3, 1965

[2] ISO 3509-1984 Glossary of terms relating to coffee and its products,

Nevertheless, these norms are the result of an attempt of standardization and stem from trade agreements and heuristic rules. Moreover, though setting precise modes of operation for the classification of defects, these standards conceive an operator who, for all his experience, may be subject to variations and imprecision in his personal perception in detecting defects.

This has spurred research and formulation of proposals as the one in [3], for a method based on the assignment of a quality rating based on gravimetric values of beans, or in [4] which underlines the relationship between the colorimetric-morphologic and organoleptic characteristics of classes of the main defects of green coffee.

These proposals have brought a windfall in knowledge both on classification methodology and on electronic sorting technology of green coffee on a colorimetric base, as sorting has become a necessity because of the quality deterioration of the raw material. In the field of sorting, in particular, improvements have been obtained in the modes of use and in the technology of these machines through a joint illycaffe'-Sortex effort that has led to an increase of patents in this domain. The following patents have proved to be most useful references: USP 4,807,762 (Gunson's Sortex Ltd, illycaffe' 1989) "Procedure for Sorting a Granular Material and a Machine for Executing the Procedure" by Illy et al. which presents a system of colorimetric sorting of coffee by means of computer assisted mapping techniques, and USP 4,699,273 (Gunson's Sortex Ltd, illycaffe' 1987) "Sorting machine" using l.e.d. by Suggi-Liverani et al. in which the optical system employs lighting units and solid-state backgrounds that perform the recognition of the size of the bean under examination.

2. Methodology employed at illycaffe'

The intrinsic variability of the characteristics of green coffees makes it difficult to maintain the quality of the final product within a narrow range. The first difficulty arises in defining the features of the raw materials and the mode of measurement.

At illycaffe', careful attention is devoted to quality control of raw material; this has led to the development and consolidation of proprietary methodology for controlling green coffee lots entering the production process.

The raw material control methodology currently consists in three steps:
• dimensional and colorimetric classification,
• olfactive classification of defective beans,
• organoleptic analysis of cups prepared from the roasted, ground and brewed sample.

ISO 4072-1982 Method of sampling green coffee in bags,
ISO 4150-1980 Green coffee: size analysis by manual sieving,
ISO 4149-1980 Green coffee: olfactory, visual examination and determination of foreign matter and defects,
ISO 6667 Green coffee: determination of proportion of insect-damaged beans.

The objective of the latter step is to assign a grade to taste, aroma and tactile characteristics (body and astringency) of the sample of the lot under examination.
At the end of the process a quality score is associated to the green coffe lot under examination, and this information is successively processed by the purchasing office.

3. Automation of the green coffee evaluation

The idea that led to the development of the present tool originates from a series of problems encountered by illycaffe' in recent years in attempting to define a sole, objective and reproducible method capable of assigning the lots a precise classification and that also takes into account the actual influence of various defects on the organoleptic qualities of the product. This paper describes an automatic tool developed at illycaffe's laboratory that solves and consolidates the manned colorimetric and dimensional classification performed in the control methodology.
The system has the following functionalities:
• automatic recognition of colorimetric and dimensional characteristics of single green coffee beans,
• automatic classification of coffee lots according to an original criterion or according to international standards,
• storage of the information on analyzed samples into a database for further elaboration.
Though not yet covering all modes of classification exhaustively, this tool permits a reduction of time and costs, and an improvement in reliability and reproducibility of measurements.
Further, by consolidating a method into the form of an instrument, this tool sets a standard and easy way of operating and does not require highly specialized personnel.

4. System's architecture

The system's architecture mainly consists of four parts:
A. sample's system of presentation,
B. optical box,
C. acquisition and control subsystem,
D. processing software.

4.1 System of presentation

Problems in reading color information of a solid opaque object are solved in

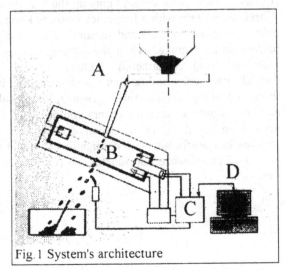

Fig.1 System's architecture

spectrophotometry by measuring the object's reflectivity. When a rapid reading of a falling object presented to a measurement system is required, problems concerning geometry and scatter are compounded by problems concerning size, fall trajectory and, in the particular case of coffee beans, asymmetry of the object and lack of uniformity of color shades.

The problem of bean presentation has been solved by a feeding system consisting in a tank containing the beans; this system is equipped with an opening from which a small quantity of material is discharged at a constant rate on a disk rotating at a constant speed. The disk is fitted with a containment spiral having the function of separating and spacing the beans. Once the bean is spaced from the successive ones by means of the spiral, it is conveyed to the rim of the disk from which it falls through a short tube into the optical box.

4.2 Optical Box

The optical box consists of three elements: lighting unit, background and sensor. The lighting unit is obtained from high intensity LEDs in red and green color bands chosen after the screening of various coffee beans by using spectrophotometric measurements. The LEDs are fitted radially on a support according to different angles to illuminate a central area in the optical box uniformly and isotropically.

Backgrounds are composed of red LEDs linearly spaced on a plate of semitransparent material and are opposed to their respective optical sensors.

The system of observation consists of three optical sensors placed at 120 degrees on an ideal circumference perpendicular to the trajectory of the fall of the bean. The sensors of the optical systems are constituted by fast silicon photodiodes.

4.3 Acquisition and Control

An electronic control circuit lights up the backgrounds and the lighting unit according to a 180 degree lag pattern with a frequency of some kilohertz. Thus, when the background is lit up, the lighting unit is turned off and sensors gather a light signal that is proportional to the degree of background coverage. When the lighting unit is on, instead, only the light diffused by the examined bean is gathered. Signals coming from the sensors are then amplified and demodulated synchronously with the lighting unit/background control system so that at the end three red color signals, three green color signals and three size signals are produced. In addition, a data acquisition system composed of a personal computer and an analog to digital acquisition board, digitalizes signals so they can be processed by the software. The next step consists in transforming the color information of the coffee bean in the optical box into two luminance/crominance coordinates, more intuitive for the operators, obtained from the color signals.

This transformation[3] entails an initial calculation of the reflectivity that is obtained by dividing the gathered color signal value by the respective size signal value. Hence, the acquired color value is independent of the size of the object. Then, the average color of the object is calculated in each of the selected color bands. Finally, color information is transformed in luminance and crominance coordinates to upgrade its manageableness.

4.4 Software

The software has been subdivided in different programs according to the functions it has to perform, instead of in a sole procedure. Given the complexity of the problem, it was necessary to resort to various systems of development in the generation of the software according to the specific problem faced in each case: a macro assembler was employed for data acquisition and hardware control, C compiler and graphical libraries for development of the other parts. One program controls the functioning capacity of the hardware and permits the acquisition of a predetermined number of coffee beans, calculates the color and size values of each bean and memorizes the results on a file. Another program processes the files containing the values of homogeneous bean groups belonging to coffee types and characteristic defects acquired previously, and generates reference maps.

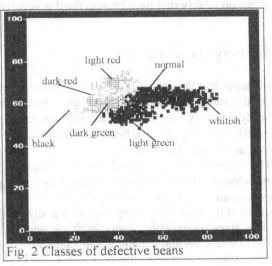

Fig. 2 Classes of defective beans

---

[3] L1   value of color signal sample at a determined wavelength acquired from optical sensor
D1   dimension of the sample of object
n   number of samples for each bean
k   compensation factor

$$R1 \quad \frac{\sum_{i=1}^{n} \frac{L1_i}{D1_i} * k}{n}$$ reflectivity of object @ L1

R2   reflectivity of object @ L2

L = (R1 + R2) / 2   pseudo luminance
C = R1 - R2   pseudo crominance

Still another program reads the file containing the data of a sample previously acquired and calculates the characteristic statistical parameters of the sample by using pre-prepared reference maps; moreover, it permits their graphic visualisation.

Thus summarizing, the logical steps performed are:

• the drawing up of reference maps by acquiring homogeneous groups of defective beans,

• acquisition of a representative sample of the lot under examination and calculation of the characteristic parameters (color and dimension),

• classification and counting of defective beans,

• sample classification according to parameters calculated in above points 2) and 3) by employing the NY standard tables or tables taking into consideration the defect's organoleptic effect on quality that have been defined at illycaffe'.

5. Fuzzy classification

In designing the classificator, the application of traditional statistical criteria does not yield good results, discriminant analysis gives bad repeatibility performances and big training sets are needed. This is mainly due to the fact that many classes overlap and it is also difficult to define exactly with experts if a bean belongs to one class or to another; so a "fuzzy" classification technique was employed .

The classificator is based on a simple assumption. Each defective class has an associated distribution profile in the L-C space; this space is superimposed on a matrix of a given granularity where each element identifies a class and a membership value. For some classes it is possible to build a class membership profile starting from frequency distribution, for others (black beans) it is very difficult to find in nature a representative sample, so we decided to merge frequency distribution with background knowledge and filled the membership profile using a matrix editor.

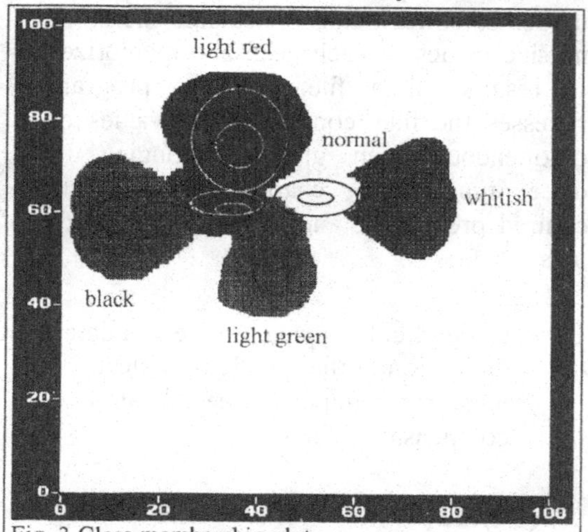

Once this classifier is built for each class of defect, we enter into the classification phase

Fig. 3 Class membership plot

by presenting to the system a sample (for example 1000 beans) to be classified, one acquired bean at a time. The bean is entered in the classifier, the class membership [0,1] is retrived, accumulated and then normalized to total number of beans, so as to obtain the percentage of each class of defect. NY index is then simply calculated by computing the number of "defects" using a predefined defect weight table and then identifying the range in the NY grid.

6. Results

The main goal in developing the system was to achieve automation in the phase of green coffee grading; in detail what has been achieved is:
• automatic generation of commodity scores,
• reduced training of operators,
• no subjective evaluation,
• reduced time evaluation (1: 30),
• higher repeatibility of evaluation,
• new functions (ex. in comparing different coffee lots).

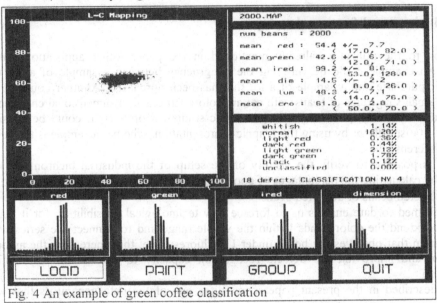

Fig. 4 An example of green coffee classification

To reach these results, the following functional objectives have been implemented:
• constant and reproducible color and size reference to eliminate problems related to personal judgments and assessment errors by personnel,
• development of a compact system equipped with a graphic workstation and control software, memorization and processing of acquired data,
• building an automatic feeder of samples with a presentation rate of 20 beans per second,
• minimization of errors due to presentation geometry (positioning, rotation),
• minimization of errors due to bean size and its natural asymmetry,
• color measurement in relation to a constant color reference comparable to spectrophotometric standards,
• measurement of bean size directly related to the sieve,

• development of pattern recognition techniques that permit the automatic classification of defects and types of coffees,
• automatic generation of report and statistical information on the acquired data also with graphical visualization.

Measurement error of the same bean is quite good; this error is mainly due to the bean's position when falling into the optical box; moreover, objects with the same color shades but with different dimensions show a comparable measurement error, thus confirming the validity of the size compensation method adopted. Bean classification by "fuzzy" method is extremely effective and, from a computational point of view, extremely efficient.

7. Conclusions and further development

In addition to the utilization already mentioned in the paper, other applications are also possible. An example could be verifying the congruency between a sample of a lot that is offered for purchase and a sample of a lot that has been purchased. Another could be is the sorting of small subgroups of beans with certain colorimetric and/or dimensional characteristics to be used in chemical, physical and organoleptic studies. Moreover, it could be possible to determine, statistically or by using the graphic representation, whether a sample is the result of a mix of different lots.

Finally, it is possible to verify the validity of the setup of the industrial bichromatic sorting machines by calculating the percentages of good beans and defective beans that are found in the group of rejected beans or accepted beans.

Work performed to date enables us to foresee new technological possibilities, for it could be possible to extend the color bands within the visible range and to connect the series with an optical system that observes the beans under UV fluorescence, thus increasing the amount of information and improving the defects recognition capacity to extend the possibilities of classification.

The tool described in the present paper constitutes one of the data generators that will be integrated into an information system for assisting the purchase of raw material which is currently in an advanced phase of study at illycaffe's laboratories.

References

[1] Clarke, R.J. and Macrae, R. (Eds.): Coffee, Commercial and Technico-Legal Aspects, Elsevier Applied Science Publisher, London, 1988.
[2] Giarratano, J., Riley G.: Expert Systems Principles and Programming, PWS Kent Publishing Company, Boston, 1989.

[3] Illy, E., Ruzzier, L.: A Proposal for a new Gravimetric System for Detecting Defects in Raw Coffee Beans, in: ASIC 3 Colloquium in Trieste, Paris, 1967.
[4] Illy, E., Brumen, G., Mastropasqua, L., Maughan, W.: Study on the Characteristics and the Industrial Sorting of Defective Beans in Green Coffee Lots, in: ASIC 10th Colloquium in Salvador, Paris, 1982.
[5] Kandel, A.: Fuzzy Mathematical Techniques with Applications, Addison-Wesley Publishing Company, 1986.
[6] Maughan, W., Milo, S., Roarzi, L.: Instrumentation System for the Analysis of Coffee Beans, in: ASIC 9th Colloquium in London, Paris, 1980.
[7] Rothfos, B.: Coffee Consumption, Gordian-Max Rieck, Hamburg, 1986.
[8] Sivetz, M. and Foote, M.E.: Coffee Processing Technology vol.1 ,AVI,Westport, 1963.
[9] Suggi Liverani, F.: A tool for the classification of green coffee samples, in: ASIC 14 Colloquium in San Francisco, Paris, 1991.

LEARNING FROM NOISY MEDICAL DATA: A COMPARATIVE STUDY BASED ON A REAL DIAGNOSTIC PROBLEM

B.A. Teather

De Montford University, Leicester, U.K.

and

G. Della Riccia

University of Udine, Udine, Italy

and

D. Teather

De Montford University, Leicester, U.K.

ABSTRACT

Clinicians routinely collect extensive case histories on their patients and in certain medical domains this data may be supplemented with confirmed or "working" diagnosis obtained by patient follow-up. The possibility of using such datasets as a source of 'knowledge' for diagnostic systems has been a goal of many research studies.

This paper reports on the application of five approaches : rule induction, neural networks, statistically based diagnostic trees, Bayes discriminants and logistic models; to the construction of diagnostic aids based on a noisy, mainly categorical, medical dataset giving the clinical presentation of patients with either Multiple Sclerosis or Cerebrovascular(Vascular) Disease who have been referred for Magnetic Resonance Imaging.

The procedures investigated gave very similar results in terms of overall diagnostic performance although the 'format' of the resulting diagnostic aids was very different. The use of Multiple Correspondence Analysis as a preparatory technique, to remove noisy variables, proved very successful in identifying a smaller subset of items that were more amenable to 'automated' techniques such as neural networks/rule induction and also assisted in the selection of variables for statistical discrimination.

INTRODUCTION

Clinicians work in an environment characterised by uncertainty [1]. Data collected from this environment is noisy, often incomplete and presents many difficulties to the developers of systems to aid in medical diagnosis. The seminal paper by Ledley and Lusted [2] proposed a probabilistic approach to diagnosis based on Bayes Theorem and, in certain well defined differential diagnostic problems, with carefully selected prospectively collected databases, this approach can provide simple and remarkably accurate diagnostic aids [3,4].

There have been numerous other attempts to utilise statistical data comprising information on presenting signs and symptoms together with follow-up diagnoses. Approaches are sometimes characterised as being based in either the Artificial Intelligence (AI) School (eg rule induction, neural networks) or the Statistical School (eg discriminant analysis). This categorisation is by no means mutually exclusive and there can be much overlap between approaches, for example the so-called Bayes Discriminant appears in both the statistical literature and many expert system shells. The dichotomy that characterises the often fierce debate of the 1980's, between the protagonists of the knowledge based and statistically based approaches [5], is no longer thought to be useful and considerable benefits can be gained from a fusion of methodologies.

THE MEDICAL PROBLEM

The Medical Systems Group at De Montfort University has been collaborating for some 15 years, with the Institute of Neurology in London, on a range of research projects concerned with the design and evaluation of computer based advisor systems in radiology [6,7,8]. A five year study, part funded by the Department of Health, has sought to establish statistical databases that can be used to provide advice in relation to the acquisition and interpretation of magnetic resonance images of the head in the differential diagnosis of cerebral disease.

Magnetic Resonance Imaging (MRI) is a flexible and powerful imaging modality that can provide high quality images of many different types("sequences") and orientations. Correct choice of sequence type, dependent on the presenting symptoms of the patient ("clinical presentation") is critical for patient management. There is ample evidence that an incorrect choice of sequence type may allow certain disease processes to be misinterpreted or overlooked [9,10]. The correct interpretation of presenting symptoms, in terms of competing diagnoses that MRI may be able to resolve, is therefore critical.

Thus the problem considered here is the well recognised difficulty of differentiating between Multiple Sclerosis(MS) and Cerebrovascular disease, which can present for imaging with similar clinical histories and symptoms[11]. It is unlikely that any computer based advisor system can be constructed to provide a perfect discrimination between these diseases based solely on clinical presentation. However, it would be useful to identify those cases for which the presentation clearly indicates one disease, and then to characterise the remaining group of patients who will benefit from further investigation by MRI and for which correct sequence choice and subsequent interpretation of the resulting MR images may be critical.

The data utilised in this study comprises the clinical presentation, coded as the presence/absence of 22 symptoms/signs (Table 1) from the patient history. In addition, patients age (years) was available for use in this analysis. Follow-up has been utilised to obtain the diagnosis for each case as one of 34 diseases. The comparative analyses are concerned with differentiating between MS and Cerebrovascular disease. Our results will be compared to an earlier analysis of this data which utilised neural networks trained by back propagation [12].

1. Loss or lowering of conscious level
2. Fits
3. Headache
4. Suspicion of raised intracranial pressure
5. History of subarachnoid haemorrhage
6. Possible meningitis
7. Clear history of head injury
8. Query demyelination
9. Cerebral (possibly single) focal features
10. Cerebellar and/or brain stem features
11. Cranial nerve signs
12. Multifocal in brain
13. Multifocal in brain and spine
14. Onset
15. Multiple episodes
16. Developmental delay
17. Chiasmal signs
18. Cavernous sinus orbital apex signs
19. Endocrine signs
20. Proptosis
21. Visual loss
22. Psychiatric disturbance

Table 1 Clinical Presentation Features

METHODS

Of the 689 cases with diagnoses, 137 were Multiple Sclerosis and 53 Cerebrovascular. This total of 190 cases was divided, at random, into a training sample comprising 127 cases and a separate test sample of 63 cases i.e. a 2:1 split (Table 2).

	Training Sample	Test Sample
Multiple Sclerosis	96	41
Cerebrovascular	31	22
Total	127	63

Table 2 Data for construction and testing of diagnostic aids

Preliminary tabulation of the training sample revealed that only 10 of the 22 binary variables relating to clinical presentation contained sufficient cases with positive signs in these two diseases to be of value in further analysis. Table 3 summarises the variables used in this comparative study. Chi squared tests of independence of feature variable and diagnostic category were computed to provide an initial indication as to which features might be informative in diagnosis.

Variable	Chi-squared	p-value
Age (continuous)	Not applicable	
Age (>50 years)	20.280	<0.001
Headache	0.055	0.815
Query demyelination	14.423	<0.001
Cerebral focal features	3.549	0.060
Cerebellar/Brain stem features	1.514	0.219
Cranial nerve signs	0.118	0.731
Multifocal in brain	1.334	0.248
Multifocal in brain and spine	13.511	<0.001
Multiple episodes	0.392	0.531
Visual loss	3.384	0.066
Psychiatric disturbance	6.293	0.012

Table 3 Variables used in the Comparative Study

Diagnostic aids were constructed from the training sample using the following methods:-

(1)   Bayes discriminant assuming conditional independence of signs (BAYES)

(2)   Linear logistic model fitted using GLIM (LOGISTIC)

(3)   Diagnostic tree constructed statistically using sequential selection of signs according to maximum chi-squared statistic (TREE)

(4)   Rule induction tree constructed using XTran (RULE INDUCTION)

The resulting diagnostic aids were then applied to the test set to compare the efficacy of the resulting procedures in terms of diagnostic error rates. These, in turn, are also compared to the earlier neural network results.

RESULTS

A critical aspect in the development of diagnostic aids is that of variable selection. There is a tendency in medical studies to record all information that is available in case it may be of value in diagnosis and decision making. This often leads to very noisy datasets in that many of the variables recorded provide little or no information for differential diagnosis.

In the case of the Bayes Discriminant, the conditional independence assumption implies that we need only examine the two way table for each feature variable against disease to judge whether or not the feature is diagnostically important. Table 3 presents the chisquared statistics for each two way table. This suggests that the distribution of features headache, cerebellar/brain stem features, cranial nerve signs, multifocal in brain, multiple episodes may be assumed independent of disease and therefore carry no information for differential diagnosis. The remaining variable subset has been included in the Bayes discriminant indicated as 'significant variables only' in Table 4. It should be noted that this model and approach to analysis assume that there are no high order interactions in the data so that a particular combination of items of diagnostic significance will be missed

The logistic discriminant seeks to model the log-odds for a disease in terms of a linear combination of features. This approach enables us to deal with a mixture of both binary and continuous data. Thus in this application we can model the increase in risk of cerebral vascular disease as the patients age increases. Standard approaches to variable selection in log-linear modelling[13] were used, leading to a model based on age(years), demyelination and multifocal in brain and spine. The overall performance of the model is similar to that of the Bayes discriminant but the diagnostic probabilities are less extreme.

Statistically based approaches to the construction of diagnostic trees have been examined for over 20 years[14,15]. A common method used for tree construction is that of sequential variable selection based on the chi-squared test statistic. No independence assumption is required and the resulting diagnostic aid can be simple to operate (Figure 1) but tends to be based on relatively few feature variables - the successive partitioning of the data results in sparse two way frequency tables so that the additional variables do not lead to a *statistically significant* increase in diagnostic performance.

In the case of neural networks and rule induction, two approaches to variable selection are possible; 'pruning' the diagnostic aid constructed from the full variable set or statistical pre-processing to reduce the number of variables submitted for analysis. Gregson[12] utilised all variables to construct a neural network for this data and subsequently pruned the network to yield a 10-2-1 architecture giving performance indicated in Table 4.

Simple application of rule induction to the 11 feature variables given in Table 3 resulted in a complicated tree with numerous branches and diagnoses at the endpoints based often on only single cases of each disease. Such a tree 'fits the training data well but gives very unreliable probability predictions about individual patients. We have already noted that approaches to variable selection based on examination of each variable singularly will miss high order associations/informative patterns in the data. This is clearly not satisfactory since one benefit of neural nets and rule induction is their ability to exploit such feature patterns. What therefore is required is a statistical pre-processing method which allows us to examine jointly the relationships between feature variables. Multiple Correspondence Analysis[16,17] provides a means of exploring the multivariate categorical data to identify redundant variables. Utilising the SIMCA software[18] we identified three features likely to provide little additional information, namely cerebellar and/or brain stem signs, multiple episodes and psychiatric. A dataset based on the reduced set of eight variables was then subjected to rule induction. Part of the resulting diagnostic tree is displayed in Figure 2 and its performance on the test set recorded in Table 4. This procedure yields the best performance in terms of overall diagnostic accuracy.

DISCUSSION

The data considered in this study is typical of many generated in medicine. The variables recorded are mainly categorical, noisy and unlikely to provide perfect discrimination between the diseases being considered.

In terms of overall diagnostic accuracy the procedures investigated all perform at a similar level, although the format of the resulting diagnostic aids is quite different. Modelling 'uncertainty' remains a critical aspect of the computer aided approach to diagnosis. The need to provide well calibrated probability statements about individual cases is well recognised. The end-user clinician needs to be able to interpret a percentage probability statement in terms of likely diagnostic error rate if informative decisions about management are to be made, since these often involve a balancing of risks. If diagnostic aids are constructed using all variables routinely recorded, the probability statements produced tend to be over-optimistic. Variable selection is therefore a key step in the modelling process. Simple examination of each variable separately against disease can be very misleading. We have found that utilising MCA as a technique for exploring multivariate categorical data is valuable as a pre-processing aid prior to the application of more formal Artificial Intelligence or Statistical Analysis techniques.

Method	Results	Accuracy
NEURAL NETWORK	6 errors 15 unclassified	$\frac{42}{63}$ 67%
BAYES - all variables	13 errors	$\frac{50}{63}$ 79%
- sig. variables only	13 errors	$\frac{50}{63}$ 79%
GLIM (LOGISTIC) - age discrete	13 errors	$\frac{50}{63}$ 79%
- age continuous	12 errors	$\frac{51}{63}$ 81%
DIAGNOSTIC TREE - 1	13 errors	$\frac{50}{63}$ 79%
- 2	7 errors 13 unclassified	$\frac{43}{63}$ 68%
RULE INDUCTION TREE	9 errors	$\frac{54}{63}$ 86%

NEURAL NETWORKS
  - pruned : $p<0.33$ predict VASC. $p>0.67$ predict MS. $0.33<p<0.67$ unclassified

DIAGNOSTIC TREES
    - 1 : $p<0.50$ predict VASC. $p>0.50$ predict MS
    - 2 : $p<0.33$ predict VASC. $p>0.67$ predict MS. $0.33<p<0.67$ unclassified

Table 4 Results of the Comparative Analyses

REFERENCES

1. Balla JI, Elstein AS and Christensen C : Obstacles to acceptance of clinical decision analysis, BMJ, 298 (1989), 579-82.

2. Ledley RS and Lusted LB : Reasoning foundations of medical diagnosis, Science, 130 (1959), 9-21.

3. de Dombal FT, Leaper DJ, Staniland JR, McCann AP and Horrocks JC : Human and computer-aided diagnosis of abdominal pain; further report with emphasis on performance of clinicians, BMJ, 1 (1972), 376-80.

4. de Dombal FT, Clamp S, Softly A, Unwin B and Staniland JR : Prediction of individual patient prognosis value of computer aided systems, Medical Decision Making, 6 (1986), 1, 18-22.

5. Spiegelhalter DJ and Knill-Jones RP : Statistical and knowledge based applications to clinical decision support systems with an application in gastroenterology, JRSS, 147 (1984),

6. Wills KM, du Boulay GH and Teather D : Initial findings in the computer aided diagnosis of cerebral tumours using CT scan results, Br.J.Radiol., 54 (1981), 948-952.

7. Teather D, Morton BA, du Boulay GH, Wills KM, Plummer D and Innocent PR : Computer assistance for CT scan interpretation and cerebral disease diagnosis, Statistics in Medicine, 4 (1985), 311-315.

8. du Boulay GH, Field B, Teather BA, Teather D and Plummer D : The extraction of expert knowledge for MR image acquisition from the published literature, Rivista di Neuroradiologia, 5 (1992), 473-482.

9. Enzmann DR and O'Donohue J : Optimising MR imaging for detecting small tumours in the cerebellopontine angle and internal auditory canal, AJNR, 8 (1987), 99-106.

10. Pojunas KW, Danials DL, Williams AL, Haughton VM   MR imaging of prolactin secreting micro-adenomas, AJNR, 7 (1986), 209-213

11. Ormerod IEC, Miller DH, McDonald WI et al : The role of NMR imaging in the assessment of multiple sclerosis and isolated neurological lesions - a quantitative study, Brain, 110 (1987), 1579-1616.

12. Gregson M, John R, Teather BA and Thompson R : Practical issues in the application of neural networks to the differential diagnosis of brain disease, Proc. IEE International Conference on Neural Networks and Expert Systems in Medicine and Healthcare, Plymouth (1994).

13. Goodman LA : The analysis of multidimensional contingency tables - stepwise procedures and direct estimation methods for building models for multiple classifications, Technometrics, 13 (1971), 33-61.
14. Teather D : Diagnosis - methods and analysis, Bulletin of IMA, 10 (1974), 37-41.

15. Sturt E : Computerised construction in Fortran of a discriminant function for categorical data, Applied Statistics, 30 (1981), 213-222.

16. Benzécri JP et al : L'Analyse des données. Tome 1 - La taxinomie. Tome 2 - L'Analyse des correspondances, Dunod, Paris.

17. Greenacre MJ : Correspondence analysis in practice, Academic Press, London 1993

18. Greenacre MJ : SimCA Version 2 - Personal Computer Software for Correspondence Analysis, User Manual, 1990.

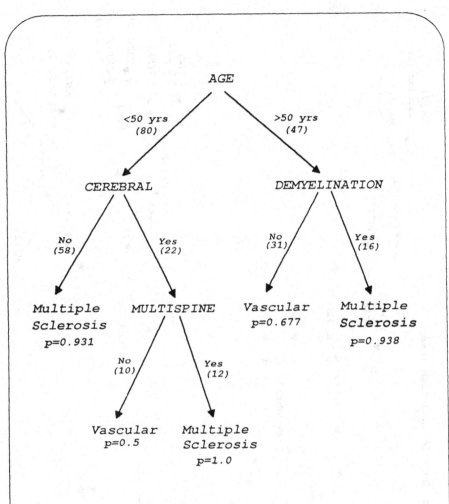

Figure 1 Chi-squared Diagnostic Tree

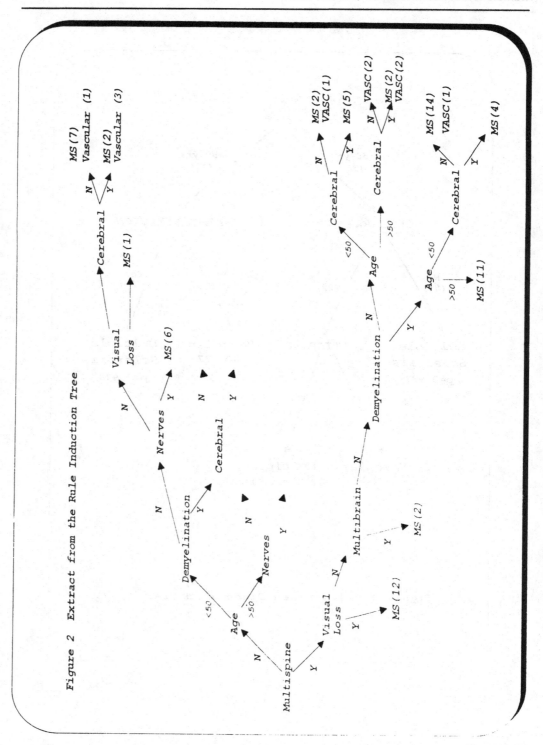

Figure 2   Extract from the Rule Induction Tree

Printed in the United States
By Bookmasters